Cultural Difference in Television Programs

Zhuo Feng

# Cultural Difference in Television Programs

## Foreign Television Programs in China

PETER LANG
Frankfurt am Main · Berlin · Bern · Bruxelles · New York · Oxford · Wien

**Bibliographic Information published by the Deutsche Nationalbibliothek**
The Deutsche Nationalbibliothek lists this publication in the Deutsche Nationalbibliografie; detailed bibliographic data is available in the internet at http://dnb.d-nb.de.

Zugl.: Jena, Univ., Diss., 2012

Cover Design:
© Olaf Gloeckler, Atelier Platen, Friedberg

27
ISBN 978-3-631-63860-6



www.peterlang.de

# Preface

This work was accepted as a doctoral dissertation at the Friedrich-Schiller-University Jena in Germany in June 2012. *Culture* is a complex whole which includes values, beliefs, knowledge, morals and habits created over time and shared by members of a society. Cultural difference varies in different ways in people's lives. Television programs also have cultural difference if they are produced in one culture and broadcast in another.

The starting point of this study concerns some controversial theoretical discussions about cultural difference in universal media products. On the one hand, there is a belief in the global cultural convergence of universal media products. In this world, people genuinely understand each other. One significant result of cultural convergence is cultural imperialism. On the other hand, the global flow of universal media products by no means indicates universal reception and the end of difference. Despite similar appearances on the surface, real differences remain and will probably stay that way for quite some time. The essence of the discussions concerns the extent of cultural difference in universal media products.

This study developed a three-stage methodological approach to examine the following three research questions from the perspective of Chinese viewers.

(1) *What types of foreign television programs are broadcast in China?* To answer this question the television schedules from 37 television channels in five sample cities (Beijing, Shanghai, Guangzhou, Nanjing, and Xi'An) in China were analyzed and an overview was constructed of the types of foreign programs broadcast.

(2) *What types of cultural difference exist in the minds of Chinese television viewers?* In this part, personal interviews with 36 Chinese viewers were conducted in five cities. The results of the interviews firstly identified various cultural-difference dimensions, which were summarized in a catalogue in order to build an overview of cultural differences. Secondly, the cultural background of each dimension was explained.

(3) *To what extent could Chinese television viewers perceive and accept cultural difference? How large is the impact of influential factors?* To answer this question a survey of 450 viewers was conducted across five cities. First, the degree of cultural difference was quantitatively evaluated based on a perception/acceptance-matrix model, which described the intercultural adaptation process of television viewers. *High vs. low perception* indicates the degree to which viewers' own cultural frame is consistent with other cultural frames in foreign television programs. *High vs. low acceptance* indicates the extent to which

viewers see the realities from other cultures as valid. Both perception and acceptance degrees combined, the stages of cultural adaptation may vary across five areas: escapees, adherents, neutralists, learners, and adjusters. Second, the impact of four groups of influential factors on perception and acceptance of cultural difference was examined, namely demographics, cultural exposure, cultural proximity, and resident city.

The results indicated that in China today, foreign television programs from the United States, Europe, and other Asian countries are a significant part of program schedules. However, most of them must be culturally modified for Chinese audiences, especially in regards to language modification.

Chinese television viewers still perceived a great deal of broad cultural differences. However, most had an open mind towards learning about cultural difference and accepted it. Although Chinese viewers still drew a clear boundary between "their" and "our" cultures, they had begun to try to see other cultures as rational and to integrate them into their own cultural framework. To what extent they perceived and accepted cultural difference was related to certain influential factors such as education, knowledge about foreign culture, cultural proximity, and resident city.

# Acknowledgements

I would like to thank Prof. Dr. Wolfgang Seufert at Friedrich Schiller University in Jena (Germany), for his supervision and for always being within reach when advice was needed. My sincere thanks go to him for his very valuable and helpful suggestions for how to conduct my research project. I also thank him for writing the recommendations for my application for scholarship. Without them, my two research residences in China to conduct interviews and surveys would not have been possible. Moreover, I thank him for encouraging me to present my research project at international conferences. I must also thank Prof. Dr. Wolfgang Frindte at Friedrich Schiller University in Jena (Germany) for his review as well as helpful comments and suggestions at the institute's colloquium.

My sincere thanks also go to Prof. Dr. Andreas Hepp at Bremen University (Germany) and Dr. Manfred Kops at Cologne University (Germany) for discussing my first drafts. I am especially very grateful to all of the interview and survey participants from the five sample cities in China. Thanks also to Pete Walton for correcting grammatical mistakes at the final stage of completing my doctoral thesis.

I gratefully acknowledge the Foundation of German Business (Stiftung der Deutschen Wirtschaft) for making my research residences in China and my participation at international conferences possible.

Many thanks go to my husband Feng Peng for providing emotional support. Above all, I wish to express my special gratitude to my parents for their emotional support in helping me finish my dissertation in Germany.

July 2012 *Dr. Zhuo Feng*

# Table of Contents

# List of Figures

# List of Tables

# Abbreviations

| | |
|---|---|
| AI | Acceptance Index |
| BJ | Beijing |
| CCD | Cultural-Difference Dimension |
| CCTV | China Central Television |
| CPC | Communist Party of China |
| CVSC | Central Viewer Survey and Consulting Center |
| CSM | China Sofres Media |
| DI | Distance Index |
| GATS | General Agreement on Trade in Services |
| GZ | Guangzhou |
| MRFT | Ministry of Radio, Film and Television |
| NJ | Nanjing |
| PI | Perception Index |
| PRC | People's Republic of China |
| SARFT | State Administration of Radio, Film and Television |
| SH | Shanghai |
| TAI | Total Acceptance Index |
| TDI | Total Distance Index |
| TPI | Total Perception Index |
| TV | Television |
| WTO | World Trade Organization |
| XA | Xi'An |

# Part A: Introduction

# 1 Introduction

## 1.1 Cultural Difference in Television Programs by Globalization

Media globalization refers to the organization, distribution and consumption of media products in the global market, particularly movies, television programs and music products. (Devereux, 2007, p. 44) Based on the recognition of the advantages of scale, media corporations target the shared habits and tastes of particular audience segments at *global level*, rather than different audience segments at *national level* on the basis of geographical proximity. (Corner, 1991, p. 28; Morley & Robins, 1997, p. 112; Keane, 2007, p. 24) Although audience segments at global level have shared habits and tastes, cultural differences still provide crucial barriers to the globalization of television programs. (Morley & Robins, 1997, p. 1)

On the one hand, television programs, which are produced for global market, must be based on "universal principles" to achieve largest number of audiences. (Morley & Robins, 1997, p. 11) These universal programs create a new global cultural space and a borderless world in which the link between culture and territory becomes blurred. Audiences transcend cultural difference between societies and understand each other mutually. (Morley & Robins, 1997, p. 112) Thus, there is a belief in world cultural convergence, in other words, a "shared culture" and a common "world awareness" among audience segments across the world. (Corner, 1991, p. 28)

On the other hand, although efforts are made to transcend cultural difference in the production of universal television programs, it is not possible to eradicate it. Despite on-the-surface similarities, real differences remain and will remain for a long time, because they are not easily changed. (Nolan, 1999, p. 1, 5) A television program rooted in one culture may have a diminished appeal elsewhere, because viewers in other cultures find it difficult to identify with the lifestyles, values, beliefs, and behavior portrayed in programs from another culture. (Hoskins et al., 1997, p. 32) Thus, universal programs do not necessarily mean universal reception and the end of difference. (Cohen, 1992, p. 32)

Some critics are worried that the cultural, social values and ideologies in media from Western developed countries (e.g., the United States and European countries) will supposedly drive out traditional cultural values in developing countries (e.g., China and India). They believe that the consumption of Western media will lead to an ideological domination by Western values; meanwhile people in developing countries will lose their original cultural identity gradually. This is known as "cultural imperialism". (Berger, 1992, p. 14)

One common question arising in these discussions is the following: To what extent has culture in developing countries really changed under the influence of Western media? In other words, to what extent can people in developing countries still perceive cultural difference in Western media? These questions are not only important for enabling media corporations to know audiences in developing countries better, but also for countries and social institutions concerned about cultural imperialism.

## 1.2 Research about Cultural Difference in Television Programs

This study aims to define and measure the changing of local culture through perception and acceptance of cultural difference in universal *television programs* from developed countries from the perspective of *Chinese audiences*. There are three reasons for this:

Firstly, a *television program* is a valuable resource for learning culture difference, because it is a universal medium with a recognizable set of cultural elements which present differences between cultures. (Mishan, 2005, p. 135) On the one hand, television as an everyday medium reflects real-life situations, such as habits, traditions, and rituals. Many of these are culturally determined and may vary from culture to culture. On the other hand, understanding the content carried on television is not only related to language, but also to a variety of cultural elements, such as values, religion, and worldviews. (Berger, 1992, p. 13)

Secondly, *China* is a suitable country for examining cultural difference. On the one hand, China is a developing country with a typical Eastern culture compared to developed countries with a Western culture. On the other hand, Chinese society has been rapidly opened up to the outside world in the last 30 years due to *China's Reform and Opening-Up Policy*. Western cultures influence Chinese culture through the media and international communication. Thus, Chinese audiences have gradually changed their opinions about cultural difference.

Thirdly, *audiences*' opinions are more important for examining cultural differences compared to the opinions of media corporations and social institutions, because using television programs occurs at an individual level. The most important research issues in this study consist of four areas:

*(1) What types of universal TV programs are broadcast in China?*

This question aims to construct an overview of the types of universal television programs, which are broadcast in China, but produced by foreign countries, including the United States, countries from Europe (e.g., England and Germany), and from Asia (e.g., Japan, and Korea). Program types are, for example, movies, television series, sports programs, documentaries, entertainment shows, news, and music programs which are produced for international markets.

*(2) What types of cultural difference exist in universal TV programs?*

This question aims to identify and summarize various cultural-difference dimensions within the cultural framework of Chinese audiences. Cultural difference here means the difference between background culture in universal television programs and Chinese culture, which could be perceived by Chinese audiences. Furthermore, the cultural reasons behind cultural difference will be explained.

*(3) To what extent can audiences perceive and accept cultural difference?*

This question aims to quantitatively evaluate the perception and acceptance degree of each identified cultural-difference dimension by viewers and compare them. The degree of *perception* indicates how much of a cultural difference viewers can perceive, whereas the degree of *acceptance* indicates to what extent they can accept cultural difference.

*(4) How large is the impact of influential factors?*

This question examines the impact of influential factors on the degree of perception and acceptance. The influential factors are, for example, demographic, television viewing, knowledge about foreign culture, and cultural proximity between importing and exporting countries.

## 1.3 Research Methods

Based on the research questions, the author developed a three-stage methodological approach to data collection combining qualitative and quantitative methods.

*(1) Pre-study: TV schedule analysis to gather television genres*

The purpose of this pre-study was firstly to construct an overview of types of foreign television programs; secondly, to describe and evaluate some forms of cultural modification in foreign programs which could reduce the cultural barriers for viewers; and, thirdly, to examine the cultural proximity between country of origin and China.

*(2) Method I: Personal interview to gather cultural-difference dimensions*

The aim of holding personal interviews with Chinese viewers was to identify various cultural-difference dimensions that were interpreted by Chinese viewers. All of the identified dimensions were summarized in a catalogue in order to form an overview of various cultural differences. Secondly, based on viewers' statements, the cultural background of each dimension was explained.

*(3) Method II: Survey to evaluate the degree of perception and acceptance*

The survey among Chinese viewers was firstly intended as a means to quantitatively evaluate the perception and acceptance degree of each cultural-difference dimension. All of the dimensions in the survey were derived from the results of the television schedule analysis and personal interviews. Secondly, all participants were divided into several viewer types according to their opinions about cultural difference. Thirdly, the survey also examined the relationships between the perception/acceptance degree and various influential factors.

This study uses a mixed form of qualitative and quantitative approaches, because each approach alone does not serve very well as a way to examine cultural difference – each one for a different reason. Qualitative interviews can explore cultural-difference dimensions and explain cultural background at a deep level, but they cannot examine the validity of various opinions about cultural difference, because the selection of participants is usually random. The quantitative survey approach offers an alternative; it can examine the validity of each opinion to a certain degree, but through a survey alone people cannot know the cultural background of viewers' opinions on cultural difference.

All of the research steps were conducted in five sample cities in China. Each city has different characteristics: Beijing is a political center; Shanghai is an economic center, Guangzhou is a port city for international trade; Nanjing and Xi'An are both cultural cities with long histories. These cities have different degrees of openness to the outside world, ranging from high to relatively low. Different sample cities were selected to garner possible different opinions about cultural differences and due to the possible existence of regional differences.

## 1.4 Overview of Chapters

This study consists of seven parts: (see Figure 1)

Part A (Chapter 1) contains the introduction or starting point of this study, the main research questions, and the chapter overview.

Part B (Chapter 2 to 5) comprises the theoretical framework. In this part Chapter 2 investigates various categories of cultural differences. Chapter 3 focuses on cultural difference in television programs. Chapter 4 discusses theories about the perception and acceptance of cultural difference from the audience's perspective. Chapter 5 summarizes part B and the limitations of the research.

Part C (Chapter 6 to 7) presents the methodological design of the study. In this part Chapter 6 introduces the Chinese television market and Chapter 7 describes the three research stages of this study: television schedule analysis (pre-study), personal interview (method I), and survey (method II).

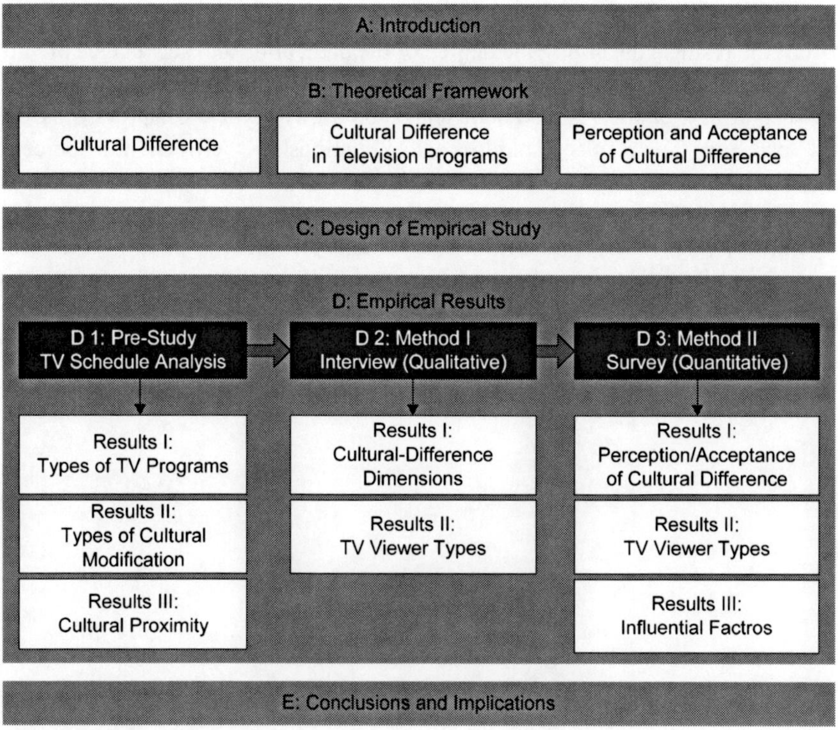


*Figure 1: Structure of Study*

Source: Own illustration

Part D (Chapter 8 to 17) presents the results of all three research stages.

- Part D1 (Chapter 8) deals with the method and results of the pre-study *television schedule analysis*. The various types of universal TV programs and types of cultural modifications are summarized.

- Part D2 (Chapter 9 to 12) presents the method and results of the *personal interviews* in order to qualitatively explore cultural-difference dimensions. Chapter 9 introduces the design and execution of interviews in the five sample cities. Chapter 10 focuses on presenting the various cultural-difference dimensions and explaining the cultural background of each dimension. Chapter 11 divides all of the participants into five viewer groups: escapees, learners, adherents, adjusters, and neutralists. Chapter 12 is the summary of this part.

- Part D3 (Chapter 13 to 17) presents the method and results of the *survey* in order to quantitatively evaluate cultural-difference dimension. Chapter 13 describes the method, analysis model, and hypothesis. Chapter 14 presents the first part of the results concerning the degree of the perception and acceptance of various cultural-difference dimensions separately and comparably. Chapter 15 is the second part of the results about television viewer types. Chapter 16 as the last part of the results examines the relationship between the perception/acceptance degrees and various influential factors. Chapter 17 summarizes this part.

Part E (Chapter 18) concludes with a summary of the theoretical framework, methods, and empirical results as well as possible implications for further research.

# Part B: Theoretical Framework

Due to globalization, in recent years many researchers have shown more interest in devising analytic frameworks to evaluate cultural difference, not only in daily life, but also in mass media. How the culture of universal media influences local culture has been an interesting research issue, because many media products are now produced for a global audience, such as films, television programs, books, and magazines.

The goal of international media studies is to gain insights into such matters as change of local culture based on the appearance of universal media culture, and related concerns such as cultural imperialism, cultural discount, and cultural barriers. These studies assumed that the domination of local values is reduced through an increasing exposure to global universal culture causing local values to become neglected.

In the theoretical framework of part B, various cultural-difference dimensions are introduced in Chapter 2. In Chapter 3 cultural differences in television programs are discussed in reference to two controversial theories – "cultural discount" and "universal media culture". Chapter 4 focuses on the "perception" and "acceptance" of cultural difference from an audience's perspective, which are two core concepts in this study. Chapter 5 summarizes the theoretical framework and some research limitations of previous studies.

# 2 Cultural Difference

## 2.1 Definition of Culture

Culture is a complex whole which includes the values, beliefs, knowledge, art, morals, customs and habits shared by members of a society. (Yeshin, 1998, p. 307) Culture is a group worldview, the way of organizing the world which a particular society has created over time. Culture decides how individuals behave and what they stand for. This framework of meaning allows members of a society to understand each other and influence the ways in which they relate to one another. Culture has three main components: "the knowledge we hold; the things we make; and the things we do." (Nolan, 1999, p. 3)

The most important precondition for understanding between members of a culture is common cultural knowledge; this is something which lies hidden inside people's heads and forms a pattern that underlies and shapes the behavior of individuals (Stewart, 1991, p. 17). "Cultural knowledge consists of the rules, categories, assumptions, definitions, and judgments that people use to classify and interpret the world around them." (Nolan, 1999, p. 3) Through this knowledge functioning as a lens through which to view their culture, people selectively perceive, organize and make judgments about the things they select. Although all of these components seem arbitrary for individuals outside a society, to the members within it these cultural rules do not seem arbitrary at all, but rather logical, normal, right, proper, and consistent. (Nolan, 1999, p. 4)

Culture is not innate, but rather comprises cultural patterns learned over a long period. The patterns developed within a social group over generations of interaction enable its members to generate meaning and structure very quickly from daily events and occurrences. (Nolan, 1999, p. 4) These cultural patterns provide them the implicit and largely unconscious cultural frames of reference to distinguish between "we" and "they". This often forms a difficult barrier for outsiders to cross because they tend to observe other people in the context of their own cultural patterns. Therefore, a foreign culture is very much akin to a secret code for outsiders, until such a time they are able to decipher it (Nolan, 1999, p. 5).

Any two cultures differ to some extent. This could be stated in terms of the degree of similarity or overlap within sets of norms defining each culture. Such similarities or differences among cultures may be roughly described along single dimensions or aspects of several cultures, for example, in terms of a common or similar history, language, religion, and education system. However, they are difficult to interpret one-dimensionally. (Torbiörn, 1988, p. 170)

Culture across the world is marked by its diversity and thus cultural differences appear in many different ways along a set of key dimensions: (1) Perceptions: People from different cultures do not necessarily see the same things, even when they are looking in the same direction. (2) Interpretations: People do not select, interpret, or remember what they see in the same way, even when they see the same thing. (3) Facts: Due to people from different cultures having different definitions of a situation, they use different pieces of information in their thinking. (4) Goals: People from different cultures may have quite different purposes or destinations in mind. (5) Values: People from different cultures apply different standards in their evaluations of individuals, situations, behaviors and outcomes. (Nolan, 1999, pp. 5–6)

In previous studies numerous scholars developed various frameworks to categorize cultural differences, primarily according to two directions – *horizontal dimensions* and *vertical levels*.

## 2.2 Horizontal Dimensions of Cultural Difference

In the categories of horizontal dimensions, each cultural-difference dimension has two opposite extreme situations. “Individualism” vs. “collectivism” as an example: American culture has the highest degree of individualism, while Japanese culture has the highest degree of collectivism. Other cultures, such as those in Europe, are located on a scale somewhere between the two poles. In some studies the horizontal dimensions were used to examine differences and similarities between particular cultures.

### 2.2.1 Hofstede’s Cultural Dimensions

One of the most famous approaches to distinguish cultural difference is Hofstede’s five cultural-difference dimensions which could affect people’s behavior in their workplace. The five cultural dimensions are power distance, uncertainty avoidance, individualism, masculinity, and term orientation (Hofstede, 2001, p. xix; Ajami & Goddard, 2006, p. 213). (See Table 1)

*Table 1: Hofstede’s Five Cultural-Difference Dimensions*

| **Cultural Dimensions** | **A Range of Cultural Variations** ←→ | |
|---|---|---|
| Authoritarianism | Low power distance | High power distance |
| Risk aversiveness | Uncertainty tolerance | Uncertainty avoidance |
| Individualism | Individualism | Collectivism |
| Attitude towards gender | Femininity | Masculinity |
| Time orientation | Short-term orientation | Long-term orientation |

Source: adapted from Hofstede, 2001, pp. xix-xx.

*Power distance* means to what extent the power of organization members is unequally distributed. *Uncertainty avoidance* is the extent to which a culture programs its members to feel either uncomfortable or comfortable in an unstructured situation. Unstructured situations are novel, unknown, surprising, and different from the norm. *Individualism* on the one side versus its polar opposite *collectivism* is the degree to which individuals look after themselves or remain integrated in groups, usually around family. *Masculinity* versus its opposite *femininity* refers to the distribution of emotional roles between genders; it contrasts “tough” masculine with “tender” feminine societies. *Long-term* versus *short-*

*term orientation* refers to the extent a culture programs its members to accept delayed gratification of their material, social, and emotional needs. (Hofstede, 2001, p. xix; Samovar, 2009, pp. 198-207)

These five dimensions are important not only in the workplace, but also in societies. For example, Chinese are used to being integrated in groups not only in work but also in private life, such as gatherings with family members and friends (collectivism). In most Chinese families many important decisions are made by men (masculinity). Furthermore, young people in China must respect old people and obey their instructions (power distance).

### 2.2.2 Kluckhohn's Basis Cultural Orientation Dimensions

Another approach to categorizing cultural difference derives from anthropologist Kluckhohn. Unlike Hofstede's five cultural dimensions in the workplace, Kluckhohn looked at culture in relation to some of life's universals. He developed five cultural dimensions and suggested how cultures might vary accordingly. (Samovar, 2009, p. 207; Salmon, 2010, p. 40) (See Table 2)

*Table 2: Kluckhohn's Basic Cultural Orientations Dimensions*

| **Cultural Dimensions** | **A Range of Cultural Variations** ←→ | | |
|---|---|---|---|
| Innate human nature | Evil | Neutral | Good |
| Relationship to nature | Subjugation | Harmony | Mastery |
| Time orientation | Past | Present | Future |
| Human activity | Being | Growing | Doing |
| Relational orientation | Dominating | Group-oriented | Individualistic |

Source: adapted from Kluckhohn & Strodtbeck, 1973, pp. 11-12.

The first dimension answers the question as to what the character is of *innate human nature*. Different cultures see the nature of humans as good, evil, or neutral. The second dimension concerns *humanity's relationship to the nature*. It examines the potential types of relations between humans and nature: subservient to, in harmony with, or mastery over. The third dimension *time orientation* explains the temporal focus of human life and examines how cultures come to terms with the past, present, and future (Schmidt, 2007, p. 28). The fourth dimension *modality of human activity* stresses three orientations of a human's primary mode of activity: stress on identity (being), stress on self-development (growing), and stress on action (doing). The final dimension *rational orientation* identifies three potential ways whereby humans might define their relationships with others: dominating, group-oriented or individualistic. (Kluckhohn & Strodtbeck, 1973, pp. 11-12; Samovar, 2009, pp. 208-214; Salmon, 2010, pp. 41-42) Individualistic is characterized by individual autonomy and preference for individual interests over group interests.

### 2.2.3 Parsons' Cultural Pattern Variables

The socialist Parsons' concept of cultural pattern variables is another view of cultural difference. The pattern variables are six dichotomous decision options, which describe the ways people have learned to meaningfully organize their experiences and perceive the phenomena around them (Schmidt, 2007, p. 30). This approach builds up a binary scheme of action as a useful instrument for explaining the role of an individual in a social situation, on the one hand, and for analyzing the basic structure of a society, on the other. (See Table 3)

The first three pattern variables are concerned with motivational orientation or value orientation. The last three pattern variables are primarily related to social objects. In each dichotomous orientation, people must choose how they define a situation. In daily life individuals rarely act at either of the two poles; they usually choose an intermediate path.

(1) *Affectivity vs. affective neutrality* orientation is concerned with what extent people are looking forward to immediate or delayed gratification and whether they show their feelings plainly by laughing, smiling or keep them controlled and subdued. (Parsons, 2001, p. xvi)

(2) *Self vs. collective* orientation concerns how people emphasize either an individual's achievement and responsibility, or the group's achievement and joint responsibility. People must choose whether to give primacy to their own individual moral standards of own individual system or social system (Parsons, 2001, p. xvi).

(3) *Universalism vs. particularism* orientation relates to equality in regard to how people are treated. People must choose whether they accept universal (universalism) or special (particularism) standards of groups. Universalism means people's behavior is oriented strongly to individual needs, whereas particularism means people behave according to social standards. (Parsons, 2001, pp. xvi-xvii)

(4) *Diffuseness vs. specificity* orientation refers to people's perceptions of public and private space. (Parsons, 2001, p. 186)

(5) *Ascription vs. achievement* orientation focuses on whether societies ascribe people's status by virtue of age, class, gender, education, race, or accord

recognition to people on the basis of what they have achieved through their own efforts or personal accomplishments (Parsons, 2001, p. 174).

(6) *Instrumental vs. expressive* orientation is related to the question of whether relational interactions with others are principally viewed as a means to other goals or are valued as an end in themselves (Parsons, 2001, p. 157; Schmidt, 2007, p. 30).

*Table 3: Parsons' Cultural Pattern Variables*

| **Cultural Dimensions** | **A Range of Cultural Variations** ←→ | |
|---|---|---|
| Gratification | Affectivity | Affective neutrality |
| Responsibility | Self | Collective |
| Treating people | Universalism | Particularism |
| Perception of space | Diffuseness | Specificity |
| Social status | Ascription | Achievement |
| Relational interaction | Instrumental | Expressive |

Source: adapted from Parsons, 2001, pp. xvi-xvii, 157, 174, 186.

## 2.2.4 Core Cultural Values

The three approaches above for categorizing cultural difference are abstract and did not use any specific cultures as examples to explain cultural difference. The approach "core cultural values dimensions" is more concrete. It uses American culture as an example culture, on the one hand, and other contrasting cultures, on the other hand.

People being able to understand what their own core values are can make them more aware of values that stand in contrast to other people's. "Although each society has many sets of values that help to make up its particular culture, some values stand out, because they seem somehow fundamental to that culture's worldview, and because they seem to organize, determine, or at least underlie, much observed behavior". (Nolan, 1999, p. 8)

Every culture has a set of core values. For example, Americans tend to be direct, open, and they believe in an individual's right to freedom, privacy, equality, and self-expression. In contrast, Chinese tend to be indirect and conservative. They avoid conflict and maintain harmony as much as possible. Although these core values may not be shared by everyone living in a particular country, they are held by a large number of people in this country. Table 4 sets out some commonly held American core values or assumptions, together with a hypothetical contrasting value.

*Table 4: Core Cultural Values*

| **American Core Value** ← | → **Contrasting Core Value** |
|---|---|
| People are individuals. | People are integrally connected with other people (e.g., family). |
| Personal growth and change are valuable and desirable. | Following time-tested traditions is desirable. |
| Individuals have control over their own life circumstances. | One's life circumstances are determined externally. |
| Personal problems can be solved through analysis and action. Others can help you with problems. | Problems are fated to occur. One's problems are beyond the control of other human beings. |
| People can be genuinely interested in the welfare of strangers. | Only one's close friends and relatives can be trusted. |
| An open discussion of one's problems can be beneficial. | It can be dangerous to reveal oneself to others. |
| People - both men and women - are more or less equal. | There is a hierarchical ranking of people in society. Males are superior. |
| Self-help and initiative determine who you are and how you live. | Inheritance or birthright is important for who you are and what your circumstances are. |
| Competition is healthy and productive. | Cooperation is the most important thing. |
| Actions are oriented toward the future, with a view to change and improvement. | There is a past or present orientation with emphasis on stability. |
| Emphasis in relationships is on directness, openness, and honesty. | Emphasis in relationships is on indirectness, ritual, "face", and smooth relationships on the surface. |
| Materialism and acquisitiveness are emphasized. | Spiritualism and detachment are emphasized. |

Source: adapted from Nolan, 1999, p. 9.

Through the core value approach the American culture was treated as the "center" of all cultures. To explain this approach in more detail, the following sections compare the core values of American and Chinese culture. Among various core values "individualism vs. collectivism" and "freedom vs. restriction" are the most significant. These contrasting core values are important for understanding this study, because they are the essential reasons why a Chinese audience can feel some cultural differences in foreign television programs.

*(1) Individualism vs. collectivism*

Individualism vs. collectivism concerns to what extent individuals are integrated into a group. Hofstede defined it as follows: "Individualism stands for a society in which the ties between individuals are loose; everyone is expected to look after himself or herself and his or her immediate family only", whereas "collectivism stands for a society in which people from birth onwards are integrated into strong, cohesive in-groups, which throughout people's lifetime continue to protect them in change for unquestioning loyalty" (Hofstede, 1991, pp. 260-261).

Americans pride themselves on being fiercely individualistic. They want to be "their own person". They are much more concerned about their own careers and their personal success than about the welfare of an organization or group. For Americans, it is every man for himself. (Hall, 1990, p. 147)

In contrast, Chinese are integrally connected with other people and are concerned more about the groups to which they belong. These groups consist of themselves and their parents, children, and relatives as a family group; themselves and their colleagues as a work group; and themselves and their friends as a friend group. In the Chinese traditional value system people consider the health and security of family members to be more important than their own health and security. Furthermore, they consider the interests and welfare of their companies and colleagues as being more important than their own. (Niemeier, 1997, p. 283)

*(2) Freedom vs. restriction*

Americans' strong awareness of individualism is directly tied to the value they place on freedom in all things, not just political freedom to vote as they wish, religious freedom, freedom of assembly, or freedom of the press, but a vast array of individual rights protected by the *American Constitution* and the *Bill of Rights*. In addition, the high degree of freedom in the United States extends to all aspects of an individual's behavior in their own life circumstances. (Hall, 1990, p. 147) Americans expect freedom of choice in virtually everything and also like to deal with their own problems by themselves. They have a great choice in terms of careers and lifestyles. They are free to marry the person of their choice with little or no interference from their parents; they are free to choose whatever job or profession they wish; and free to attend the college of their choice if they are accepted and can afford it (Hall, 1990, p. 147).

Chinese collective culture is directly tied to the value the Chinese place on harmony in everything, for example, harmony in daily life, in the workplace, and between family members, friends, relatives, and colleagues. A consequence of this is that the life circumstances of someone who is Chinese are determined externally. Chinese pay a great deal of attention to the feelings of other people. Thus, their choices are often influenced by their parents, children, relatives, friends, and so on, when making decisions, such as whom to marry, in which college to study, and where to work. Their choices are not only important for themselves, but also for these related persons.

Although some Chinese desire freedom of choice, they simply cannot have it, because they live in a collective society with a lot of personal restrictions. If they always do everything according to their own wishes and neglect the feelings of other group members, they will be considered an "outsider" of a group and lose the trust of other people. Chinese usually only trust close friends and relatives (Hillebrand, 2008, p. 7), because they think it can be dangerous to reveal themselves to strangers. That is the reason why personal relationships (*Guanxi*) are very important everywhere in China.

*(3) Directness vs. indirectness*

The preferred level of directness among people from different cultures varies. (Barker, 2002, p. 249) American and Chinese also have different preferences for directness and indirectness in interpersonal communication.

Americans believe that relationships should be direct and open. They pride themselves on directness, because directness implies honesty and trust and is a very desirable characteristic to display.

In contrast, Chinese emphases indirectness, ritual, "face", and some smooth relationships on the surface. Indirectness is part of hospitality in Chinese culture. In China people believe that truth is relative to circumstances and human obligations. Therefore, they prefer to tell each other what they believe that person wants to hear instead of the absolute truth (Barker, 2002, p. 249). Chinese believe that one should not criticize other people, especially those in authority, in order to avoid other people losing "face". If they want to inform other people about problems or make suggestions to them, they prefer an indirect way. "The true lie" is sometimes used to make other people feel better.

*(4) Improvement vs. stability*

Americans' actions are oriented toward the future, with a view to change and improvement. In contrast, in China there is a past or present orientation with emphasis on stability. In a country as highly populated as China, stability (*Wen Ding*) whether political, social or economic has often been of great concern. Chinese are traditionally afraid of social instability. In the imperial period, population pressure had destabilizing effects upon Chinese society, and sometimes led to large-scale rebellions (Tubilewicz, 2006, p. 2). Thus, the traditional three teachings (*San Jiao*), i.e. Confucianism, Daoism, and Buddhism favor stability and harmony (Tian, 2004, p. 28; Wang, 2010, p. xiv). In contemporary China, social stability is also the most important precondition for development.

*(5) Femininity vs. masculinity*

In America both men and women are more or less equal, whereas in China males are traditionally seen as superior. (Brownell, 2002, p. 2) China is an agricultural country. In the imperial period males, were the main labor force in each family. Without males in the family, people could not get enough food to survive. At that time, only males were considered as intelligent and allowed to sit the state exam for a civil-service position. The main task of females historically in China was to take care of their children, husband, and other family members. However, in contemporary China inequality between males and females is not as obvious as it was in the past.

## 2.2.5 Hall's High- and Low-Context Culture

### 2.2.5.1 High- and Low-Context Culture

In the four approaches mentioned above to categorizing cultural-difference dimensions, all of the dimensions have two opposite poles and each particular culture is positioned somewhere in between. A limitation of these approaches is that they only summarize a few key dimensions. The number of cultural-difference dimensions is actually huge. Therefore, people need a framework for categorizing a huge number of cultural dimensions in a simple manner.

One of the most important theoretical concepts is "high- and low-context culture" elaborated by anthropologist Edward T. Hall. High- and low-context culture indicates how much information is viewed as enough in a culture. Cultures differ across a continuum that ranges from low to high context.

Firstly, two concepts related to high- and low-context should be distinguished: *Code* is the explicit, formulated content of communication, whereas *context* is the situation and relational aspect, which includes implicit, unformulated rules governing how information is handled and how people interact. (Gozzi, 1992, p. 56)

Hall used context as a verb; it means *to context*, in other words, to integrate information into a cultural pattern. *Context* is the information that surrounds an event and information is inextricably bound up with the meaning of that event. The elements that combine to produce a given meaning are in different proportions depending on the culture. (Hall, 1990, p. 6)

The context surrounding any message may be relatively dense or relatively sparse, which bears consequences for the style of communication. Certain cultures, which Hall called "high-context cultures", typically have denser contexts surrounding their messages; others typically have fewer contexts and are called "low-context cultures". (Gozzi, 1992, pp. 55-56) "A *high context* communication or message is one in which most of the information is already in the person, while very little is in the coded, explicit, transmitted part of the message. A *low context* communication is just the opposite; i.e., the mass of the information is vested in the explicit code." (Hall, 1990, p. 6)

There are some differences between high- and low-context cultures:

*(1) Implicit (indirect) vs. explicit (direct) style of communication*

On the one hand, information and rules are implicit in high-context cultures that use an indirect style of communication (Schmidt, 2007, p. 24; Hillebrand, 2008, p. 6). They are typically short, pithy, and poetic. They can only be understood by a member of the "in-group" and remain somewhat puzzling to outsiders, who do not glean enough information from them. (Ajami & Goddard, 2006, p. 213) When dealing with touchy issues, high-context communication is not specific about them, but instead weaves a web of inference around the issue, which, it is assumed, both parties are aware of and can draw conclusions from. Low-context intruders often see this as indirection and untrustworthiness (Hall, 1977, p. 113). Therefore, people in high-context cultures tend to be extremely reserved, with much more being taken for granted and assumed to be shared, thus permitting an emphasis on understatement and nonverbal codes.

On the other hand, information and rules are explicit in low-context cultures that use linear logic and a direct style of communication. They must be longer, elaborated more, and more explicit than high-context culture. (Ajami & Goddard, 2006, p. 212) Touchy issues are dealt with directly, openly, which high-context viewers seen as impolite and clumsy. (Gozzi, 1992, p. 57)

*(2) Closed vs. open systems of thought*

High-context cultures tend toward "closed systems" of thought, which are self-validating and discourage fundamental questioning. In high-context cultures the sign for something is seen as closely connected with its referent. At its extreme, this is "magical" thinking, where the manipulation of a sign is thought to affect the thing the sign represents.

Low-context cultures, in contrast, tend toward "open systems" of thought, more readily admitting indeterminacy, and more often questioning the received cultural patterns. In low-context cultures the sign is thought of as only arbitrarily connected to its referent. (Gozzi, 1992, p. 57)

*(3) Collective vs. individual personal relationship*

In high-context cultures commitment between people is very strong. Responsibility to others takes precedence over responsibility to oneself. (Keegan & Schlegelmilch, 2001, p. 81) High context is not a given; it is an achievement, constantly being constructed. To maintain a high context for communication people must spend a great deal of time together, talk with each other, and pay attention to each other. (Gozzi, 1992, p. 58)

In low-context cultures the emphasis is placed on individuals, with the bonds between people being more tenuous and the extent of involvement and commitment to long-term relationships being lower. (Keegan & Schlegelmilch, 2001, p.81) In low-context cultures, people spend less time together and more time alone. (Gozzi, 1992, p. 58)

The differences introduced above between high- and low-context cultures can virtually affect every situation and every relationship in which the members of these two opposing traditions find themselves. High-context people are apt to become impatient and irritated when low-context people insist on giving them information they do not need. Conversely, low-context people are at a loss when high-context people do not provide enough information. One of great communication challenges is finding the appropriate level of contexting needed in each situation. Too much information leads people to feel they are being talked down to; too little information can mystify them or make them feel left out. Ordinarily, people automatically make these adjustments in their own country, but in other countries their messages frequently miss the target. (Hall, 1990, p. 9)

Each particular culture has different needs for contexting – the process of filling in background data. Hall distinguished particular cultures within a range between two extreme poles of "high context" and "low context" on a scale. (Son, 2007, p. 19) (See Figure 2)

*High-context cultures* are Japanese, Chinese, and Arab people involved in close personal relationships and have extensive information networks among family, friends, colleagues, and clients. Thus, they usually neither require, nor expect much in-depth background information through communication. This is because they keep themselves informed about everything having to do with people who are important in their lives.

*Low-context cultures* include Scandinavians, Germans, English-Canadians, and Americans. They compartmentalize their personal relationships, work, and many aspects of daily life. Consequently, each time they interact with others they need detailed background information. (Hall, 1990, p. 6) The French, Italians, and Spanish need fewer contexts than the Japanese and Chinese, but they need more contexts than Germans and Scandinavians.

Based on the continuum between high- and low-context cultures, cultural proximity between cultures can be examined. For example, Chinese culture and German culture have low cultural proximity, whereas Chinese culture and Japanese culture have high cultural proximity. In this study *cultural proximity* is a very useful concept for classifying numerous cultures into a few groups and thus making the comparison between cultures much easier.

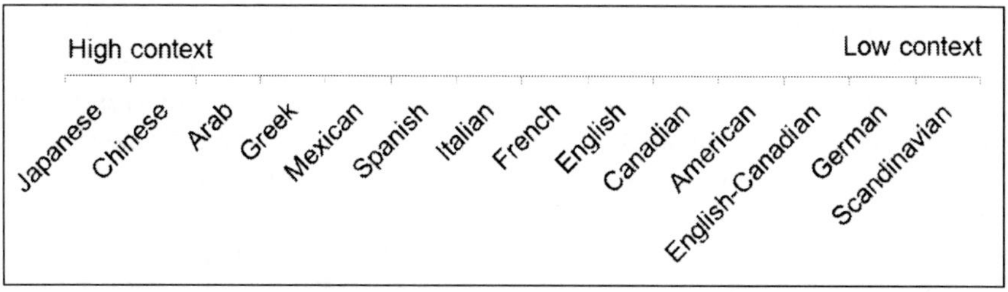


*Figure 2: Continuum between High- and Low-Context Cultures*

Source: adapted from Krizen, 2007, p. 36.

### 2.2.5.2 High- and Low-Context Culture in Mass Media

High- and low-context culture also has an influence on mass-media content. In high-context cultures such as in Japan and China, the mass media is generally reliant on contextual cures. This means that many of messages are implicitly expressed in context instead of in a lot of words. (Gozzi, 1992, p. 55) Thus, in high-context countries television programs tend to be more intuitive and appeal to viewers' emotions.

In contrast, in low-context cultures such as Germany and America the mass media place most of their emphasis on the explicit, coded part of message. This means that many of the messages to be communicated are expressed explicitly in words and there is little need to take account of surrounding circumstances. (Pickton, 2010, p. 70) In low-context cultures, people have relatively little shared experience that can be called upon by using mass media. Mass media must often provide a background of shared stories, symbols, and characters that people will have in common. (Gozzi, 1992, p. 62) Thus, programming in low-context cultures trends to be more logical, scientific and provides evidence to substantiate the content and opinions in program.

Due to the differences of mass media in high- and low-context cultures, television programs produced in low-context culture must provide fewer illustrations and less information if they are broadcast in a high-context culture. In contrast, high-context television programs must provide more illustrations and information if they are to be broadcast in low-context culture. If mass media have to fit into both high- and low-context culture, they move to the middle of the continuum instead of remaining near the high- or low-context ends. (Gozzi, 1992, p. 58) Specifically, mass media in high-context cultures will tend to move more toward the low-context end of the continuum, whereas in a low-context culture mass media tend to move more toward high-context end. (Gozzi, 1992, p. 59) This middle position is occupied by *universal media culture*, such as the culture of Hollywood movies which avoid high- or low-context factors as much as possible.

This tendency of mass media moving from lower context towards higher context may lead to some predictions. The high-context culture's closed cultural systems will be relativized by intrusions of universal mass media. This makes people in high-context cultures less able to confer meanings and leads to a reduction of the domination of local cultural systems. Gradually, people have shown a greater willingness to use more universal media products. (Gozzi, 1992, p. 61)

This is often called cultural imperialism or media imperialism. Media products from Western developed countries, especially movies and television programs, undermine the indigenous culture of a less developed society. Importing Western media products also indicates importing Western values and encouraging people to envy and seek to become part of the alien culture they are viewing. (Chaffee, 1992, p. 37) Therefore, universal mass media products from Western countries may push a high-context culture more towards the low-context end of the spectrum.

## 2.3 Vertical Levels of Cultural Difference

When using cultural-difference *dimensions* as an approach to distinguish cultural difference, all dimensions are considered as being equal and the relationships between them ignored. In this chapter cultural-difference *levels* consider the hierarchical relationships between them and cultural difference *layers* distinguish them in interrelated layers from inside to outside. Both concepts identify principal elements and coordinating systems composing a culture.

### 2.3.1 Levels of Cultural Difference

Culture is often compared to an iceberg – much of it lies beneath the surface, out of people's immediate awareness. People generally respond to the surface values that they can sense; however, to truly understand a culture, they must also explore the behaviors below the waterline. This is a useful metaphor for examining the technical, formal, and informal levels of culture. (Schmidt, 2007, p. 22; Hatim, 2004, p. 162)

(1) *Technical level* is the explicitly clear and visible portion of the cultural iceberg and includes artistic, technological, and materialistic components of a culture as well as its institutional systems. (Hamilton, 2010, p. 124) People generally find the fewest intercultural misunderstandings or problems at this level. (Schmidt, 2007, p. 22)

(2) *Formal level* of the cultural iceberg rests at sea level, partially above and slightly below surface. It includes norms, rules, roles, traditions, rituals, customs, and communication patterns of a group. Norms provide a guide for how group members should behave, and rules clarify what is mutually considered "right" and "wrong" by a group. (Hamilton, 2010, p. 124) Roles define and explain expected or actual performance in relationships or social situations, whereas traditions, rituals, and customs describe regular practices within a culture. (Schmidt, 2007, p. 23)

(3) *Informal level* of cultural iceberg extends far below sea level and includes cultural history, cultural identity, core values and beliefs that shape a culture's worldview and influence cultural identity. The cultural history and background of a culture can generate insights into the norms of a group and assist people in understanding a culture's identity. (Hamilton, 2010, p. 124) Cultural identity is people's sense of who they think they are. It commonly emerges through daily social practices and reflects the values as well as the worldview of a culture. (Martin & Nakayama, 2009, p. 213) Core values convey what is good and bad, as well as express what is proper and improper, or what behavior is normal and abnormal. A culture's worldview consists of a "set of more or less systematized beliefs and values in terms of which evaluates and attaches meaning to the reality that surrounds it". It is unconsciously accepted as the way things are. (Schmidt, 2007, p. 23)

## 2.3.2 Layers of Cultural Difference

The system of interconnected cultural layers is often seen as an onion. The underlying assumption of the onion is the core value of a culture, while the outside layers represent behaviors and artifacts that reflect core value. (Miller, 2008, p. 93; Drenth et al., 2001, p. 119) Culture has three distinct but interconnected layers:

(1) *Artifacts* (surface layer) comprise highly visible manifestations of culture that can be observed and identified using the senses, such as behaviors, dress

style, rituals, publications, and stories. (Watkins & Leigh, 2009, p. 345) They are easy to access, but difficult to interpret without a prior understanding of the deeper layers of culture. (Küng, 2009, pp. 174–175) In other words, people can see what is presented, but they do not know why these particular characteristics are being displayed.

(2) *Espoused values* (middle layer) are at least partially visible. They are the things that people consider as having value, which can sometimes be found in written records and spoken expressions, such as ethics, vision, and philosophies. (Watkins & Leigh, 2009, p. 345; Küng, 2009, pp. 174–175) This culture layer may appear to reveal a culture's underlying beliefs.

(3) *Basic assumptions* (deep level) are essence of culture and largely invisible. They consist of basic norms, values, beliefs, perceptions, and feelings about a culture which are developed through shared experience. They act as the ultimate source of values and drivers of actions. (Watkins & Leigh, 2009, p. 345) Most are unconscious and invisible to members of the culture and to those who try to understand it. Basic assumptions contain the key to a culture and the tools by which the other two layers, artifacts and espoused values, can be interpreted. (Drenth et al., 2001, p. 119)

Onion model is an interconnected system. However, the surface layer can be modified over time. (Yeshin, 1998, p. 307) Uncovering and deciphering a culture belief is difficult, because underlying assumptions lying at the heart of a culture operate beyond conscious awareness and they are not accessible at the surface attitudinal level. (Küng, 2009, pp. 174–175)

# 3 Cultural Difference in Television Programs

Cultural differences exist not only in daily life, but also in the mass media. For example, television programs as cultural products reflecting various cultural differences. Some are general cultural differences the same as in daily life, e.g., value difference, language difference, and difference of way of life. Some are special television cultural differences related to television genres, such as the cultural difference of characters and the drama plot of television series, the commentary style in sports programs, and the objectivity of documentaries.

## 3.1 Intercultural Television Communication

### 3.1.1 Intercultural Television Communication

Television programs are more than just a commodity; they are also cultural goods as they say something, which must be understood in a cultural context. They are culturally "read" and interpreted by television viewers (Drackllé, 2001, p. 200). Each audience uses its own memories of everyday life as a form of cultural context in order to understand a TV program (Luhmann, 2002, p. 56), which include personal experiences, expectations, attitudes, identities, knowledge, and their own beliefs.

Therefore, television as an everyday medium demands more or less cultural proximity with the lives of television viewers. (Bennett & Strange, 2011, p. 266) The feeling of cultural affinity or cultural difference of television programs depends not only on the use of language, but also on a range of national differences, such as value orientations, aesthetic tastes, lifestyles, traditions, clothing, rituals, music and narrative style.

To understand cultural differences in television programs, people should firstly know what role cultural differences play in the communication process of international TV programs.

Figure 3 is adapted from Shannon and Weaver's model of communication. It presents a communication process from television production to acceptance by audiences. Each television producer encodes and sends messages, and each television viewer receives and decodes them in turn. Producers encode the information they want to transmit to viewers in television programs. Television programs are then transmitted through television channels to viewers. Finally, viewers decode information in their own cultural context and give a response to producers.

The communication process of television program is not defined as a container of messages processed in a line from sender to receiver, but as the means through which people construct meaningful cultural elements by which to live. (Burnett, 1996, pp. 65–66) The encoding and decoding devices are primarily cultural, and involve a diverse set of elements, including words, gestures, and symbols, whose meanings are essentially arbitrary. (Nolan, 1999, p. 34) The "cultural context of the TV producer" encompasses the whole environment in which TV producers create meanings (Sklan, 1996, p. 236) and it influences television content they produce (Rohn, 2011, p. 632). The "cultural context of the TV viewer" encompasses the whole environment in which viewers understand and accept meanings. It influences what content they select, how they interpret it, and to what extent they like it. (Rohn, 2011, p. 632)

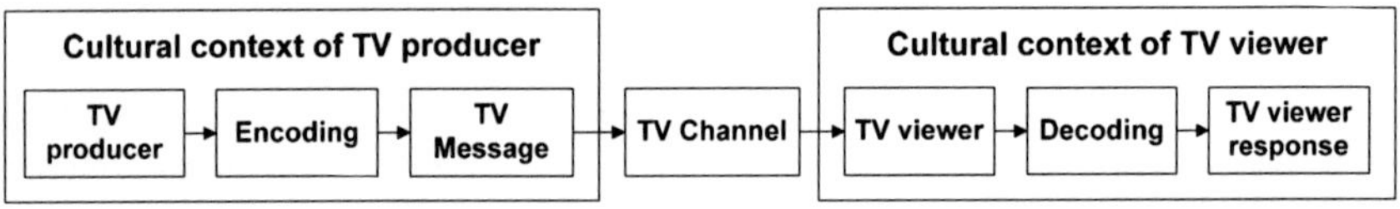


*Figure 3: Television Communication Process*

Source: adapted from Fiske, 2002, p. 7; Narula, 2006, p. 26.

Figure 4 depicts an *intercultural* television communication process. For example, the BBC produces news and documentaries which are broadcast to viewers in China. Producers from countries of origin encode and send information. Through international channels viewers from countries of destination receive and decode information. Different to Figure 3, the *intercultural* television communication process (Figure 4) occurs in two different cultures of country of origin and country of destination.

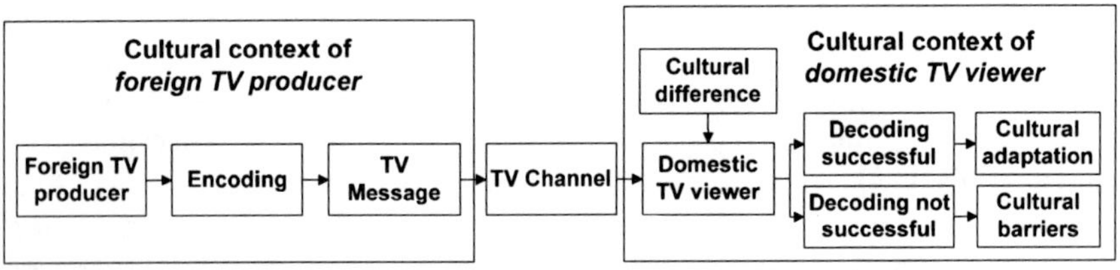


*Figure 4: Intercultural Television Communication Process*

Source: Own illustration.

### 3.1.2 Mutual Knowledge between Television Producer and Viewer

In the receiving phase, cultural differences, which are the core research object in this study, have a great impact on the reception of television programs. Reception of foreign programs is not only related to viewers' attitudes towards cultural difference, but also to viewers' knowledge about cultural difference.

(1) *Viewer's attitude* towards cultural difference: As messages in foreign television come in, viewers attempt to make judgments and evaluate them according to their own cultural framework which has taught them how to think about the world. On the one hand, viewers may let their own feelings or interpretations intrude, preventing them from finding anything to like in the foreign culture. On the other hand, they may not allow their own culture to work against something unfamiliar in foreign culture and they try to understand it.

(2) *Viewers' knowledge* about culture difference: Encoding and decoding are two opposite processes. Whether encoding and decoding are successful, depends on how much cultural background, knowledge, and experiences the producers and viewers have in common. If this shared area is very small, in other words, viewers do not have enough personal experience and knowledge to decode messages, they will encounter a variety of problems and obstacles when trying to understand foreign programs. In this situation, cultural difference is considered as a cultural barrier. A consequence may be that the viewer does not want to watch the program. In contrast, if this shared area is large, it means producers and viewers have enough knowledge and experiences in common to decode the message, they can more or less adapt and integrate culture difference with their own culture.

Rohn proposed a "Lacuna and Universal Model" to determine whether the understanding of media content is influenced by "cultural baggage" of media producers and media users. (Rohn, 2011, p. 632) Lacuna is a Latin word meaning "gap" which is often used in the context of misunderstanding between cultures regarding not only linguistic features, but also for values, attitudes, knowledge, experience, and expectations, that exist in one culture but not in another. (Rohn, 2010, p. 89)

In this model, the term *Lacuna* describes the phenomenon that audiences do not select and enjoy media content produced outside their own culture if they perceive a mismatch between their own cultural baggage and that of the producer as incorporated into the content. In contrast, the term *Universal* relates to the phenomenon that audiences would like to select and enjoy media content which is produced outside of their own culture if no significant mismatch of cultural baggage is present. (Rohn, 2011, p. 634)

Thus, international television program makers must not only estimate the state and size of viewers' knowledge base (Al-Shammari, 2010, p. 34), but also gain a cultural understanding of the meanings of the entire spectrum of cultural elements in programming, such as colors, symbols, and costume. This is to ensure that the design and message of their products communicate appropriate and the desired values needed to satisfy the needs of foreign viewers.

### 3.1.3 Cultural Barriers

If the producer from the country of origin did not encode appropriately and viewers from the country of destination could not decode properly, then cultural barriers emerge. "*Cultural barriers* refer to individual reactions in situations of cross-cultural contact where cultural differences negatively affect an individuals' ability, willingness to understand, accept, adhere to, or adopt the norms of a foreign culture." (Torbiörn, 1988, p. 168)

Cultural barriers represent negative rather than positive evaluations, dislike rather than like, and avoidance rather than acceptance of viewers' reactions to foreign programs, because some aspects of culture in foreign television may appear less normal, less familiar, less good than the viewers' own cultural standards. (Torbiörn, 1988, p. 171)

In addition, cultural barriers should not be manifested uniformly or to the same degree across whole range of possible reactions. This concept required specification:

- Which aspect of another culture do individuals confront? Cultural barriers may appear lower in regard to cultural difference at a superficial level (e.g., clothing, food) than cultural difference at a deep level (e.g., value, view of world).

- How central is that aspect to individuals? Any cultural difference, large or small, may constitute a high or even "absolute" barrier if it touches on a fundamental and central aspect of an individual's own culture.

- How long does a cultural barrier exist? A cultural barrier tends to become lower for a person as time passes, because the individual's perception of what is normal and appropriate changes. (Torbiörn, 1988, p. 172)

## 3.2 Dimensions of Cultural Difference in Television Programs

Several communication studies showed that people in different cultures and societies understand and interpret identical television content differently, such as the same scenes in news, soap operas, and movies without modifications (Cohen, 1992, p. 28; Berger, 1992, p. 15; Butler, 2007, p. 48). This is because they watch television programs through the prisms of their knowledge as well as cultural and social backgrounds, which provide them different kinds of cultural tools and frameworks. (Morley, 2002, p. 80)

For example, *Dallas* is broadcast in 90 different countries and has been studied extensively by scholars. It is popular in some Western countries, but it is not popular in some Eastern countries (e.g., Brazil and China) where it gets very low viewing figures. One of the most important reasons for this is that viewers in one country may decode *Dallas* differently from viewers in another. (Massing, 1988, p. 194; Berger, 1992, p. 15)

Television programs, which fail to appreciate the significance of local cultural factors, will fail in a foreign market. Several examples illustrate this: it would be an anathema to show pork or shellfish ingredients in programming intended for a predominantly Jewish country; the same would apply to beef in Hindu communities or alcohol for Muslims. Besides such obvious errors, the surroundings in which the message is set (a home, a retail outlet, etc.) may similarly contradict existing cultural beliefs. In some markets, for example, it would be inappropriate to depict a woman wearing Western clothes, because it is regarded as taboo. (Yeshin, 1998, p. 307)

Television program is multi-sensorial. Thus, cultural elements that cause a different understanding and interpretation of television programs are diverse. Numerous scholars analyzed various dimensions and notions of cultural difference in television programs that make intercultural comparison possible. The typical cultural differences, which result in markedly different attitudes towards foreign television programs, are language and dialogue, characters, drama plots, and values:

(1) *Language and dialogue*: Use of language and dialogue is quite different in various cultures which could indicate class, identity, values, and so on. For example, in the United States, accents in news shows, situation comedies, and soaps do not reflect socioeconomic background the way they do in England. *Humor* in language and dialogue also varies considerably from country to country, from class to class, and gender to gender. (Berger, 1992, p. 16)

(2) *Body language*: Body language, such as gestures, handshakes, kissing, and facial expressions can vary from culture to culture. Indeed, what is cultural-

ly determined can be misleading and misunderstood. Smiling and raising an eyebrow accompany greetings in many cultures. The degree of eye contact; the distance people stand from one another; whether, when, how often, and whom people touch and kiss also vary between cultures. Body language is also sometimes shown in television scenes; for example, a hand being placed over a glass to summon a waiter in a restaurant, and knife and fork being positioned in such a way as to indicate the meal is finished. (Mishan, 2005, p. 135)

(3) *Characters*: *Characters' social class and characteristics* vary in different cultures, for example, in the English series *Upstairs Downstairs*, class differences are all important, whereas American soaps tend be classless. (Berger, 1992, p. 16) *Characters' roles* (particularly gender roles), such as gender, age, race, ethnic type might be over- or underrepresented in different cultures. One example is the use of sexuality in television commercials. Commercials from Western countries tend to be much more overtly sexual than Asian ones. *Protagonists' characteristics* mean the characters of heroes and heroines, as well as villains and villainesses, whose characteristics and functions in television programs could reflect cultural differences. (Berger, 1992, p. 17)

(4) *Drama plot*: *Plot structure* is concerned with how the plot is constructed. For example, societies that are hierarchical and see authority as "legitimate" (e.g., England) tend to have solutions to story problems imposed from above, in contrast to egalitarian societies (e.g., the United States) where this is not accepted. (Berger, 1992, p. 17) *Dominant themes* of drama plot might also be different. For example, the themes of American soap operas reflect their dominant ideology that success is a function of willpower, where all people have opportunities. Therefore, heroes of these programs would be business and professional people who "prove" that their ideology is correct. In English shows, in contrast, show aristocratic types whose values are quite different from American ones. (Berger, 1992, p. 18)

(5) *Value*: It is reflected through characters and stories in television programs what are understood as general notions of good and bad, what is desirable and what is undesirable. It decides what is espoused and what is attacked, downplayed, or neglected.

(6) *Visual style*: This consists of camera work, types of dominant shots, lighting, and editing techniques. For example, the way football games are televised in America, Germany, and France is different. (Berger, 1992, p. 18)

(7) *Style of broadcasting company*: This indicates the broadcasting companies' attitudes towards audiences, such as the degree of commercialization, styles of presentation, and content balance between proportions of social problems, politics, sport, game shows, drama, arts, and so on. (Mishan, 2005, p. 135)

## 3.3 Media Culture Difference: Cultural Discount Theory

In most countries television viewers prefer to watch domestic programs rather than foreign programs when all conditions are equal. (McAnany & Wilkinson, 1996, p. 68) The programs made in one's own language and containing familiar cultural messages will be more highly valued. Thus, cultural products diminish their value in other cultures. This value diminishment is known as *cultural discount* found in economic literature addressing trade in cultural goods. (Petty et al., 2005, p. 279) The term cultural discount is closely related to the term *cultural proximity*, since people prefer programs that come from similar cultures. Culturally proximate media products experience less cultural discount. (Sora, 2004, p. 267)

### 3.3.1 Definition of Cultural Discount

Hoskins explained *cultural discount* as the following: "a particular television programme, film, or video rooted in one culture, and thus attractive in the home market where viewers share a common knowledge and way of life, will have a diminished appeal elsewhere, as viewers find it difficult to identify with the style, values, beliefs, history, myths, institutions, physical environment, and behavioral patterns." (Hoskins et al., 1997, p. 32)

Cultural discount arises primarily because viewers in destination markets generally find it difficult to identify some unfamiliar and inexperienced cultures in foreign television programs. As a result, fewer viewers watch a foreign program than a domestic program of the same type and quality. (Hoskins & Mirus, 1988, p. 500) The reduction in appeal, which is reflected in the price the broadcaster is willing to pay, is regarded as cultural discount. (McAnany & Wilkinson, 1996, p. 68)

Cultural discount attached to a given imported TV program is calculated as the value reduction in a foreign market relative to a domestic equivalent: (Hoskins et al., 1997, p. 32; Hoskins et al., 2004, pp. 47–48)

$$\text{Cultural discount} = \frac{\text{Value of domestic equivalent - Value of import}}{\text{Value of domestic equivalent}}$$

Indeed, this concept is an anti-cultural term designed to measure how cultural difference affects the market value of cultural commodities in a quantitative way. (Acland, 2003, pp. 33–34) A product with low cultural discount is more successful outside the domestic market, which is an important consideration for entertainment companies with global interests. (Vaidya, 2006, p. 401)

### 3.3.2 Television Program as Cultural Good and Economic Good

The notion of cultural discount originates from the dual nature of the cultural industries. The term *culture industries* may at first glance seem like something of a contradiction. *Culture* means the whole realm of ideas, images, symbols, objects and practices which make all of social life meaningful. And *industry* is something to do with factories and labor. It refers to almost any area of productive activity. Cultural products are the output of cultural industries. (Hoskins et al., 1997, p. 1)

Television is a relatively new form of cultural production. In fact it shares two characteristics: a business and a cultural embodiment. (Napoli, 2004, p. 209) It is an industry because its "products" are primarily made for profit and bought and sold in markets, the output of which is always ultimately concerned with the production, distribution and exchange of clusters of symbols, signs and meanings. (McIntyre, 1996, pp. 216–217)

(1) *Economic goods*: Nearly all major media companies are commercial corporations, whose primary function is generating profits for their owners and stockholders. The yardstick by which the business performance of media companies is measured by investors has little to do with public-interest concerns about creativity, independent thought, and diversity. Instead, it has everything to do with sales, advertising revenue, and profits. (Croteau & Hoynes, 2001, p. 1)

(2) *Cultural goods*: Television deals in ideals, information, and culture. It gives form to social life in words and pictures, sounds and images. It offers the terms and symbols which people live their lives by. (Burnett, 1996, p. 34) It is a cultural good specific to the values, beliefs, styles, and the like of target populations. Television produces a mental landscape, a worldview, a particular way of thinking about the reality of society. (Moran, 1996, p. 4) Television informs and entertains people, influencing how they understand themselves and their world, as well as how people spend their leisure time. Therefore, television plays a significant cultural role in societies and holds a unique position in democratic societies, because it makes free expression and independent thought possible. (Croteau & Hoynes, 2001, p. 1)

(3) *Cultural and economic goods*: Television industries are comprised of primarily symbolic goods, and as a consequence, their economic value can never be disassociated from their cultural value. It is hard to separate the cultural and

commercial characteristics of television. (Reca, 2006, p. 185) Thus, television industries face a great dilemma in attempting to reach both economic and cultural goals. When cultural and economic goals were in opposition, it appeared that culture normally won out. Culture, which "encodes an enormous variety of meanings and messages into economical and emotionally powerful forms", should be the starting point for economic strategy. (Küng, 2009, p. 173)

## 3.3.3 Influential Factors of Cultural Discount

In many studies about cultural discount the language difference was considered as a major factor in determining cultural discount. The lowest international discount factor is attached to the English language, hence the more successful penetration of English-language television programming into world markets than other languages. Although the linguistic similarities or differences seem to be a more decisive basis for comparative advantage or disadvantage (McAnany & Wilkinson, 1996, p. 44), language is far from the only factor which contributes to cultural discount (Collins, 1990, p. 191). However, not all Anglophone cultural products are equally exportable. People must ask a more specific question about which types and varieties of television programs are exportable to whom and under what conditions.

### 3.3.3.1 Television Genres

*Genre* is a French word meaning type or kind. The repertoire of elements that serves to identify genres consists of character types, setting, iconography, narrative style and so on. (Neale, 2007, p. 3) As such, it has played an important role in the study of literature, theater, film, television and other art and media forms. (Neale, 2007, p. 1)

Television genres are, for example, drama (e.g., action series, police series, hospital series, and science fiction), soap opera, comedy (e.g., sketch comedy, situation comedy, and adult animation), popular entertainment (e.g., quiz show, talk show, sports and music programming), children's television (e.g., preschool programming and children's cartoons), news, documentary, reality TV, and educational programming. (Creeber, 2007, pp. v–vi)

Cultural discount vary with television genres and thus explains why trade is concentrated in a few categories. For example, some scholars suggested that cultural discount is much greater for entertainment shows (e.g., talent show, game show, talk show) than for informational programming (e.g., channels specializing in news and current public affairs programming). And cultural discount of these informational programming is higher than dramas. (Morley & Robins, 1997, p. 63)

Cultural discount of news programs is large. Not surprisingly, very few Canadians with the ability to pick up U.S. stations watch U.S. network or local news, although the success of CNN and the BBC World Service has demonstrated that a significant global market segment is willing to watch foreign-based international news.

Within drama, action drama typically crosses cultural barriers (including those of language) relatively unscathed and it is not surprising to find that it is the most traded television program type. Musicals, westerns and gangster dramas lost less when dubbed or subtitled. (Acheson & Maule, 2005, pp. 335–336) Situation comedy does not export as well, since national tastes in humor differ and language subtleties are lost in translation. (McAnany & Wilkinson, 1996, p. 69)

### 3.3.3.2 Demographics

The degree of cultural discount is also influenced by viewer demographics. Gender, age, class, education, and other socio-demographic characteristics of television viewers in a foreign market have an impact on the extent to which the value of imported programming is discounted. There are two controversial opinions about the demographic diversity of television viewers:

*(1) Viewers are regional diverse.*

Nation-state is characterized by heterogeneity of population within its national boundaries. There are differences of class, ethnicity, language and accent, region, history and religion, to name some principal points of difference across populations. (Moran, 1996, pp. 10–11) Accordingly, within national boundaries individuals belong to multiple cultural communities with various cultural identities.

A consequence of a diverse population on the television industry is that the global audience markets are highly complicated. People select what they like from what they watch. Their tastes are not simply monolithic. The reasons for this selection are influenced by many factors, some of which have little to do with the aesthetic qualities of a particular audiovisual product, but rather social demographic characters ranging from age, race, marital status, gender, education and geographical location. (Burnett, 1996, p. 36)

For example, in northern Europe young people are the heaviest viewers of satellite channels and imported television programs. Younger viewers are, however, becoming increasingly familiar with foreign-language material. Older viewers are more likely to avoid imported programs for language and cultural reasons. Similarly, in Holland the high entertainment output of Sky and Super Channel is attracting a young Dutch audience. (Morley & Robins, 1997, p. 61)

Besides age, cultural discount is also related to social class. Acceptance of foreign-language material is predictably highest among the better educated, those in the higher social groups and those living nearer the capital. (Morley & Robins, 1997, p. 61) Viewers, who want to watch many kinds of imported programs, tend to be concentrated in the middle and upper classes. In contrast, the imported programs might be said to suffer a higher cultural discount among those outside the middle and upper classes. (Bondebjerg & Golding, 2004, p. 69)

*(2) Viewers are not regional diverse anymore.*

Attributed to the technical advances of telecommunication, lifestyle differences between particular societies are less pronounced. As the global consumption capital is built up, the local culture may play a diminished role in television consumption (Ginsburgh & Throsby, 2006, p. 1082). The traditional market segmentation approach by demographics such as age and location is no longer as practical with the rise of information-based attitude groups that share similar consumption patterns. For example, the New Age media consumers from Chicago and Tokyo may have more in common with each other than the same consumer in Chicago may have with a person from a nearby suburb. This parallelism provides further incentives for globalization of media companies and their respective media products. (Albarran & Chan-Olmsted, 1998, p. 12)

#### 3.3.3.3 Cultural Proximity

Clearly cultural discount varies depending on the cultural proximity between producers and consumers. (Collins, 1990, p. 212) The term *cultural proximity* refers to the fact that audiences in general have a stronger preference for cultural products originating from their own cultural background and reflecting their own language, history, and values. If given the choice, most of them prefer TV programs that are nationally and/or locally produced. For example, in Europe alone, U.S.-made television programs account for less than 3% of primetime programming and less than 1% worldwide. (Gershon, 2006, p. 224) That is the reason why international TV programs without customization for different national markets have in most cases not been a success.

The trade in television programs follows normal trading patterns in the sense that closer cultures trade more with each other. The positive effects of common language and geographical closeness are significantly stronger for the trade in television programs than for overall trade. The larger the cultural distance between export and import countries, the higher the cultural discount. (Fu & Sim, 2010, pp. 121–122; Ginsburgh & Throsby, 2006, p. 1082) Cultural dissimilarities diminish the importer's receptivity to exporter's programs and they receive fewer programs from a country more culturally disagreeable.

The notion cultural proximity is closely related to television viewers' cultural preference, because the degree of proximity is finally adjudged by self-selected viewers. They have a free choice of which program, and how many units of each, they watch. (Hoskins et al., 2004, p. 78) Viewers' preference decides the interplay of supply and demand of television programs and fundamentally influences the answers to three great economic questions: *what to produce, how to produce, and for whom to produce.* (Albarran & Chan-Olmsted, 1998, pp. 6–7) To sell television programs internationally, media companies must select the messages that viewers in other cultures prefer. (Küng, 2009, p. 7)

*(1) High cultural proximity inside region*

The growth of regional television markets, where producers based in a country within the region dominate trade, is because cultural discount is less for foreign programs from within a region than for foreign programs from outside a region. These regional markets are described as *cultural-linguistic markets* in which audiences share the same or similar languages, histories, and broadly overlapping cultural characteristics. (Hoskins et al., 1997, p. 35)

This partially explains why Egypt has become a primary exporter in the Arab world, and Japan and Korea play a central role as regional producers in Asian communities. (Sora, 2004, pp. 272–273) In recent years, countries like China, Vietnam, and Mongolia have readily adopted the television cultures of Japan and Korea, which is sometimes referred to as the "Japanese Wave" or "Korean Wave". (Sora, 2004, pp. 267–268)

For example, Taiwanese audiences feel greater cultural proximity with Japanese television programs than American ones, because most Japanese dramas depict the mundane lives of ordinary people rather than the glamorous dream worlds of the rich that are often depicted in American drams. For example, although the American medical drama *E.R.* (Emergency Room) is a high-quality production and is well received in Taiwan, it does not get people emotionally involved with the world of medicine. In contrast, the Japanese version of *E.R.*, *Kyumei Byoto 24ji* featured a female intern as the story's main character and emphasized the layperson's view of medicine instead of a professional practitioner's. This made it appear more intimate and ordinary than *E.R.*. (Iwabuchi, 2002, p. 149) Another Japanese drama about medical students *Kagayaku Toki no Nakade* also makes many people in Taiwan feel like working in the medical profession.

*(2) Low cultural proximity outside region*

Programs dominating in regional markets will have a highly diminished value outside the region. For example, Japanese television programs would be receive a high cultural discount in Western countries (e.g., U.S. and European), because they tend to be limited to "cultural neutral" consumer technologies for regions outside Asian communities. (Iwabuchi, 2002, p. 26) First, the Japanese language is not widely spoken outside Japan. Second, Japanese culture is quite different to Western cultures. Third, Japanese actors look foreign to Western audiences, who thus find it harder to associate with the story. Fourth, Western audiences are used to a much faster pacing with more frequent scene changes, than Japanese audiences. Furthermore, music also poses problems. (Hoskins et al., 1997, p. 33)

*(3) Low cultural proximity as advantage*

In many studies large cultural distance between origin and destination country were often regarded as a disadvantage for international trade. However, sometimes television programs from culturally distant countries are even more attractive than domestic ones, because they are exotic, different, and less boring for viewers.

Generally, cultural difference embodied in foreign programs from culturally distant countries can be seen as less of a threat to local programs precisely because they are conceived as "foreign", while those originating from culturally similar countries might be more threatening. Japanese media imperialism in Asia is a case in point. (Iwabuchi, 2002, p. 26)

- **Language Proximity**

It is difficult to find an exact system to calculate cultural discount in high-grossing television programs and consider all kinds of influential factors. However, it is clear that language dissimilarities give rise to cultural discount. (Hoskins et al., 1997, pp. 39–40) Language is a "natural" barrier to the trade of television programs and one of the most important components of cultural discount, "as the appeal of viewing is reduced by the need to dub or subtitle and by the difficulty in understanding unfamiliar accents". (Hoskins et al., 1997, p. 4) "Even if the language is the same, accents or idioms may still cause problems." (Hoskins et al., 1997, p. 32)

As a consequence, it is common for French to dub imported Quebecois programs into Parisian French. The British film *Riff Raff*, which contained a wide range of UK/Irish regional and ethnic accents, had to be shown in North American cinemas subtitled. (Hoskins et al., 1997, p. 32) The reason is that U.S. viewers are usually insular and intolerant of foreign programming. They not only do not accept dubbing or subtitling, but they are also averse to British accents. (Hoskins et al., 1997, p. 45)

Language is often the most important criterion in a host nation's decision to import foreign television programming. In Austria, for example, almost 12% of the country's television imports come from neighboring Germany. Similarly, Belgium and Switzerland are both major importers of French programming. (Gershon, 2006, pp. 224–225) Furthermore, language proximity decides how large a price discount is. The more substantial the language barrier crossed, the heavier the discount imposed; thus a U.S. television program that costs $ 20,000 in Anglophone Canada may cost only $ 200 in Central Africa. (McAnany & Wilkinson, 1996, p. 18)

The lowest international discount is attached to the English language. Hence English-language producers, such as the U.S., the U.K., and Australia, are particularly advantaged and enjoy a lower discount when accessing other English-language national markets. In addition, English is the world's principal language and hence English-language productions are often more acceptable than other foreign-language productions in non-English-language markets. (McAnany & Wilkinson, 1996, p. 72)

In an attempt to offer a more differentiated analysis, Collins argued that cultural discount is likely to be lower for works with little linguistic content than for works in which speech is an important element. (Morley & Robins, 1997, p. 62) For this reason, a television channel such as MTV builds its international appeal precisely on the fact that its content poses few problems with language, and words are practically irrelevant. Programming that is not dependent on understanding language – for example, an opera or ballet channel or a pop channel – will attract widely dispersed and sometimes potentially large audiences. (Morley & Robins, 1997, p. 63) Conversely, programming with high speech content (e.g., informational programming) is not popular with foreign viewers, because it may cause a great deal of misunderstanding and encounter language barriers.

- **Value Proximity**

Value proximity is also seen as being determined by international television program trading. First, television producers create symbols reflecting their basic values. Second, audiences select from a wide variety of television programs those that are most consistent with their values. Television programs, which do not conform to particular basic values in a culture, might be rejected by it. (Hoskins et al., 1997, p. 33)

For example, due to value dissimilarity U.S. television programs find difficulty gaining viewer acceptance in Japan. In 1981 the audience share of *Dallas* was less than 10% in Japan, while that of *Oshin* a Japanese serial drama was over 50%. The suspense in *Dallas* arises from greed, self-interest, lying and manipulative behavior that might be considered objectionable and shameful in Japanese culture that prides itself on loyalty, self-sacrifice and honoring one's obligations. (Hoskins et al., 1997, p. 33)

## 3.4 Media Culture Convergence: Universal Media Culture Theory

### 3.4.1 Definition of Universal Media Culture

Through globalization many television producers are increasingly targeting not only their own domestic market, but also a global audience. Appealing to audiences comprised of various linguistic and ethnic groups requires neutralizing the attributes of programs. Television programs which lumber themselves with cultural specificity will not sell as well as those without it. (Acland, 2003, p. 33) Therefore, the strategies of some big media corporations, especially U.S. producers, aim to create audiovisual products with a universal appeal and promote a global audiovisual space. (Goff, 2007, p. 53)

Building a global audiovisual space is not only a matter of economics and technology, but also one of culture. It is closely associated with improving mutual knowledge among audiences around the world and increasing their consciousness of the life and destiny that they have in common. Many television makers are now making an effort to produce programs which appeal to global audiences and help them to develop a sense of belonging to a universal culture community, step by step.

International financing, settings, stars and production teams breed international television programs appealing to most people in most places. Many international programs made under these conditions have become universal favorites. (Moran, 1996, p. 9) Television channels, such as *CNN*, *MTV*, and *Discovery* are seen as laying the foundations for a post-national audiovisual territory. In effect, the format and type of television programs originating in American has created a new universal art form which is claiming something close to a worldwide audience. (Hoskins et al., 1997, p. 44)

### 3.4.2 Motivation: Maximizing of Market Size

Most companies operating in television industries are commercial operations whose main motivation is profit. Economic profits in the long run are necessary and this has forced companies to maximize revenue through maximizing audience size. The value of a program is directly related to its advertising revenue.

This will be a function of the number of expected viewers and the worth of each viewer based on their income. Potential viewer numbers depends primarily on market size. Therefore, maximizing viewer numbers is the primary goal for attaining profit. (Hoskins et al., 1997, p. 73)

The television industry, which deals in culture, is faced with high levels of uncertainty, because it is difficult to forecast the success for any particular work. A high degree of risk translates into a high rate of failure. Most ideas are never turned into series. The few that are produced often lose money, and those that lose money usually do not return for a second season. (Hoskins et al., 1997, p. 51)

In such a high-risk environment the first order of business must be the development of risk-reduction strategies. Major media companies have adopted two key strategies to ensure their long-term survival. The first is a large-scope operation which permits them to maintain a diversity of projects at all times. This enables them to offset the many loss-making projects against the occasional big winner. (Hoskins et al., 1997, p. 51) The second is a large-scale strategy to maximize market size. To achieve the largest audience, many television programs, no matter where they are made, are produced with an international audience in mind. (Cohen, 1992, p. 24) For example, for an Australian television maker the United States is an important part of his or her market. (Acheson & Maule, 2005, pp. 335–336)

Co-production is a very important way to maximize market size, which could improve access to a partner's market. Firstly, a foreign partner is likely to have better knowledge regarding the distribution process in his domestic market and better connections to key players. Secondly, a foreign partner will have superior knowledge of the attributes demanded by viewers in his market and can help ensure the program possesses such attributes. (Hoskins et al., 1997, p. 104) In effect, the creative input of a partner should ensure that the final program suffers a minimal discount in the partner's market.

### 3.4.3 Result: Cultural Imperialism

The terms *media imperialism* and *cultural imperialism* are frequently used to sum up the domination of American audiovisual products in the news and entertainment fields. U.S. television has achieved success by developing global universal programming which reduces cultural discount. (Bondebjerg & Golding, 2004, p. 69) Thus, the United States has long been the dominant exporter with a net trade balance for audiovisual products of approximately $3 billion a year (McAnany & Wilkinson, 1996, p. 71).

The United States not only has a language advantage, but also some historical and cultural specificity which leads to its domination in the global media market. (Acland, 2003, p. 33) It enjoys the unique combination of a large population with a common language, cultural diversity, and the universal appeal of American media culture. (Hoskins et al., 1997, p. 38)

(1) *Large home market*: The home market in the U.S. is large and all television providers have a great opportunity for cost amortization in the home market.

(2) *English language*: The base language in the U.S. is English, the second most-used language after Mandarin Chinese, with the majority of speakers residing in the wealthiest countries. This constitutes a global competitive advantage for its television industry.

(3) *Cultural diversity*: The United States are a melting-pot society and television programming must be broadly based for diverse viewers. Therefore, many tried-and-tested television dramas that are successful in the competitive, polyglot, U.S. domestic market are generally also successful in many other foreign markets.

(4) *Universal appeal of American media culture*: The U.S. has a competitive advantage in the creation and global distribution of popular tastes because of the textual transparency of media products. *Textual transparency* is defined as any textual apparatus that allows audiences to project indigenous values, beliefs, and rituals onto imported media or their usage. This transparency effect means that American media exports manifest narrative structures that easily blend into other cultures. Those cultures are able to project their own narratives, values, myths, and meanings onto the American iconic media, making those texts resonate with the same meanings they might have if they were indigenous. (Acland, 2003, p. 33)

However, this universal media culture which is dominated by the United States is not without its problems. Not everyone feels attracted to this kind of universal culture, because they can still sense some cultural barriers to accepting universal culture. Evidence tends to show that far from accepting universal culture supposedly implicit in American media products, international audiences actively interpret the programs within their own cultural framework. (Straubhaar, 1992, p. 212)

Evidence also tends to show that cultural capital, identity and language tend to favor an audience desire for cultural proximity; which leads audiences to prefer local and national productions over those which are globalized and/or American, since the indigenous programs' familiarity renders them easier to comprehend and closer to their everyday concerns. (Straubhaar, 1992, p. 212) The presence of a production sector in local culture industries is a necessary and sufficient condition for the cultural expression of a population, because the indigenous programs functioned as a forum for introspection to consider themes of identity and current political, cultural and social issues. Imported material hardly ever raises such issues and cannot have this kind of authentic relationship with local audiences. (Moran, 1996, p. 9)

For example, in Europe there is a growing realization that the success of American commercial programming is context-dependent in a very specific sense. U.S. imports only do well when domestic television is not producing comparable programming. Whenever viewers have the alternative of comparable programming in their own language, the American programs tend to come off second best. (Morley & Robins, 1997, p. 63)

# 4 Perception and Acceptance of Cultural Difference

The common questions, which both theories of cultural discount and universal media culture are concerned, are the following: Can television viewers perceive and accept any cultural difference in universal television programs, and to what extent? In other words, how well can they adapt cultural difference in these programs?

The answers to these questions are important, because (1) they are related to the degree of cultural discount of international television programs. The less cultural difference viewers accept, the larger the cultural discount. (2) The answers to these questions could adjudge to what extent the universal television programs are really "universal" for viewers. (3) They could also examine to what extent cultural imperialism has really occurred and to what extent American media products have homogenized diverse indigenous cultures.

## 4.1 Stages of Intercultural Adaption

The term intercultural adaptation is defined as the entirety of phenomenon of individuals, who strive to adjust their minds and behavior to an unfamiliar cultural environment, establish and maintain a relatively stable and functional relationship with a new culture. (Gudykunst, 2005, p. 380) At the core of this definition is the goal of achieving an overall person-environment "fit" between their internal conditions and conditions of new culture. (Gudykunst, 2005, p. 380)

The main theoretical perspectives concerning intercultural adaptation consist of three groups: Some researchers focused on the stage model of intercultural adaptation, some elaborated the culture-learning models, and others discussed intercultural adaptation in terms of one's cultural identity. The theoretical perspective here is only concerned with stage models, because only in these models can the degree of intercultural adaptation be considered.

A number of scholars conducted various studies to observe and examine the intercultural adaptation process that results from intercultural contact and encounters. The sections below will briefly describe some of famous stage models that focus on cognitive developmental process through intercultural contact. The early models of intercultural adaptation include the *U-Curve Adaptation Model* (Lysgaard, 1955), and the *Five Stages of Transitional Experience* (Adler, 1975). More recently, Bennett proposed the *Developmental Model of Intercultural Sensitivity* (DMIS) (Bennett, 2007).

### 4.1.1 Lysgaard's U-Curve Adaptation Model

The *U-Curve* and *W-Curve* models of intercultural adaptation are two frequently cited developmental models, which adapt the concept of *cultural shock* (Furnham, 1986, p. xvi) to a stage model of cultural adjustment. Lysgaard (1955) proposed a U-curve of adjustment during cross-cultural relocation based on his empirical investigation of Scandinavians in the United States. (Ward, 2001, p. 81) The U-curve hypothesis was later expanded by Gullahorn (1963).

This model proposed that there is a multistage wave response of adjustment and satisfaction in response to the acculturation of people who are living abroad or spending a substantial amount of time in a different culture. This intercultural adaptation tends to pass through three phases of the U-shape process:

(1) In the *honeymoon* stage (the beginning high range of the U-curve) people have a high level of excitement. The initial experiences in a culture may well reveal in a honeymoon stage in which experiences are relatively positive in the content of a halo effect of novelty.

(2) In the *hostility or stressful* stage (the bottom of the U-curve) people have a rapid drop in adaptation, because the starkness of the cultural differences rudely confront a person's adaptive resources. (Earl, 1999, p. 152)

(3) In the *final* stage (another high part of the U-curve), people are expected to recover and even recognize the humorous nature of the incongruities between the cultures. They pull themselves back up into an effective phase of managing their assignments abroad and gradually recover to a balance of psychological comfort. (Stella, 1999, p. 247)

### 4.1.2 Alder's Five Stages of Transitional Experience

Alder's (1975) *Five Stages of Transitional Experience* model describes cross-cultural adaptation as a set of developmental transitions leading to greater cultural awareness: contact, disintegration, reintegration, autonomy, and independence. (Bringle, 2010, p. 251; Gonzalez, 2004, p. 22)

In the first stage *contact*, people experience excitement and are curious about the differences of a new culture. In the second stage *disintegration*, people begin to become aware of cultural differences and contrast them. They increasingly have the feeling of loneliness and isolation, because people in a new culture behave inappropriately. Distress and withdrawal may take place in this stage. In the next stage *reintegration*, people begin to strongly reject the new culture, because they are sometimes angered by, frustrated with or anxious of cultural difference. They are increased to contact with people from the same culture and rebuild their cultural identity. In the stage of *autonomy*, their intercultural sensitivity of the new culture increases and cultural differences are less perceived and emphasized. They are more confident because of their improved language and social abilities to communicate with people from the new culture. In the last stage of *independence*, people feel more comfortable in the new culture and become increasingly independent of their original culture. (Gonzalez, 2004, pp. 22-23; Pedersen, 1995, pp. 245-246)

Alder noted that culture shock in earlier stages is a necessary part of the process of people discovering their cultural identities and promoting intercultural experiences. The cultural changes may cause people to experience different degrees of positive and negative reactions. The value of this cultural experience may facilitate self-awareness of their cultural identities. In this process they gradually realize and develop self-concepts of their own cultural identities.

### 4.1.3 Bennett's Developmental Model of Intercultural Sensitivity

Bennett proposed the *Developmental Model of Intercultural Sensitivity* (DMIS) that attains a level of complexity and makes the description of people's experiences of intercultural adaptation more useful. This model links changes in cognitive structure to an evolution in attitudes and behavior towards cultural difference. (Bennett, 2007, p. 26; Easterby-Smith, 2005, p. 224)

In this model, the intercultural adaptation process is divided into ethnocentric stages and ethnorelative stages. *Ethnocentric* is defined as using one's own set of standards and customs to judge all people, often unconsciously. In contrast, *ethnorelative* refers to being comfortable with many standards and customs, as well as to having an ability to adapt behavior and judgments to a variety of interpersonal settings. (Bennett, 2007, p. 26) Short descriptions of the six stages of development are shown below. (See Figure 5)

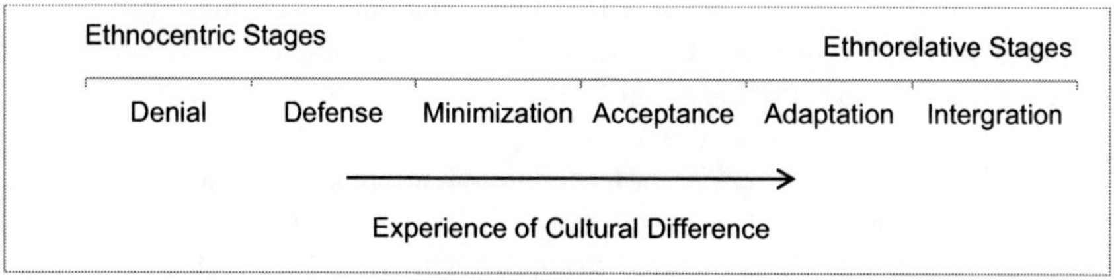


*Figure 5: Developmental Model of Intercultural Sensitivity*

Source: adapted from Bennett, 2007, p. 26.

(1) *Ethnocentric* stages consist of denial, defense, and minimization. In these stages people engage in cultural evaluation. The goodness or ethicality of actions is given by absolute and ethnocentric principles. Thus, people guide their own behavior along ethical lines according to their ethnocentric principles.

- In the *denial* stage, people are unable to construe cultural differences in complex ways, because they live in relative isolation from other cultures. Either they do not perceive cultural differences at all, or they can conceive only broad categories such as "foreigner" or "Africans". People in this stage may use stereotypes in their descriptions of others due to knowing only one or two things about other people. (Bennett, 2007, p. 26; Martinelli, 2003, p. 30)

- In the *defense* stage, people have more ability to construe cultural difference. To avoid the possible threat of change in their stable worldview, they evaluate other cultures with negative stereotypes and tend to be "underdeveloped"; contrarily they attach positive stereotypes to their own culture and view it as "developed". The defense may also have a reversed form. Some people consider their own culture as "underdeveloped" and become zealous of an adopted new culture. (Torikai, 2009, p. 170; Bennett, 2007, p. 27)

- In the *minimization* stage, people try to bury cultural differences within already-familiar categories of cultural similarity. They recognize and accept superficial cultural differences such as eating customs and other social norms, but they assume that all people are essentially the same – just human. (Bennett, 2007, p. 27)

(2) *Ethnorelative* stages consist of acceptance, adaptation, and integration. In these stages, people are inclined to interpret and evaluate behavior from a variety of cultural frames of reference, so that there is never a single right or wrong answer. (Bennett, 2007, p. 30)

- In the *acceptance* stage, people enjoy recognizing as well as exploring cultural differences. They begin to accept diverse ways of thinking and behaving in different cultures. This is the first stage in which people begin to think about cultural relativity. They understand that their own behavior and values are not the only good way to act in the world. In this stage, people have moved beyond ethnocentric rules for thinking and behavior and may not yet have developed ethnorelative principles. (Martinelli, 2003, p. 30; Bennett, 2007, p. 28)

- In the next stage of *adaptation*, people intentionally use knowledge about their own and others' cultures to shift into a different cultural frame of reference. They can understand things across cultural boundaries through taking another person's perspective. Based on their ability to use alternative cultural interpretations, people in this stage can modify their behavior in ways that make it more appropriate to cultures other than their own. They can operate effectively in more than one culture. (Bennett, 2007, p. 28)

- In the last stage *integration*, people attempt to reconcile the sometimes conflicting cultural frames that they have internalized. In a cultural transition process, some people can no longer identify with either their own original culture or adapted culture. They move into integration and see themselves as interculturalists or multiculturalists. They recognize that their worldviews are collective constructs. (Martinelli, 2003, p. 30; Bennett, 2007, p. 29)

## 4.2 Perception and Acceptance as Two Different Concepts

Almost all theories about intercultural adaptation are concerned with two questions: (1) To what extent can people perceive cultural difference? (2) To what extent can people accept cultural difference?

In numerous studies about cultural difference in media products, two concepts "perception of cultural difference" and "acceptance of cultural difference", which are used to interpret audiences' opinions about cultural difference, were not clearly distinguished from each other. Actually, they are two quite different concepts and indicate two different aspects of audiences' opinion about cultural difference.

*Perception* indicates the degree to which people's own cultural frame is consistent with another cultural frame. *Acceptance* indicates the extent to which people could accept other cultural frames and see perceptual realities from other cultures as valid. However, perception is an indicator of the size of the perceived cultural differences, not their acceptance. A large cultural difference does not indicate a difficult acceptance. For example, food culture between China and Italy may be perceived as large, but this difference is very easy for people to accept. In contrast, some cultural differences may be perceived as small, but they are difficult for people to accept.

In this chapter, the author developed a *perception-acceptance-matrix model*, in which "perception of cultural difference" and "acceptance of cultural difference" are two different dimensions for describing the stages of intercultural adaptation. Each has two degrees: high vs. low perception and high vs. low acceptance. The main purpose of simplifying various complicated stage models of intercultural adaptation into this matrix model is to make the comparison of numerous cultural-difference dimensions possible.

## 4.2.1 Perception

### 4.2.1.1 Perception of Cultural Difference

Before an explanation of the perception of cultural difference, people should first know something about human sensation. *Human sensation* is the neurological process by which people become aware of their environment which includes sight, hearing, smell, taste, and touch. (Jandt, 1998, p. 170)

People's sensations are not alike. For example, the sensation of the world of a person suffering from color blindness is very different to another person's. No two people can assume that their sensations are the same. In fact, very different stimuli can produce the same sensations, the same stimulus can produce very different sensations, and the route from stimulus to sensation is in part conditioned by culture. To some degree, cultures affect what people sense, but they have a much greater influence on human perceptions. Jandt noted that "culture has a much greater effect on the perception process than on sensation itself". (Jandt, 1998, p. 172)

Although everyone may sense in same way, cultures teach them how to process and understand the information obtained from their senses. (Jandt, 1998, pp. 185–186) *Human perception* is usually thought of as three steps which are affected by culture:

(1) S*election*: In this stage people choose what sensory information to pay attention to. Their senses constantly provide them with an enormous amount of information. They cannot be consciously aware of it all. Therefore, they must choose only the most relevant information for them. Their cultures teach them what is relevant. For example, they pay close attention to the sounds of their native language, but it is more difficult for them to hear and tell the difference between the sounds of foreign languages which they have not learned. (Jandt, 1998, pp. 185–186)

(2) *Organization*: Along with selecting stimuli from environment, people must organize them in a meaningful way. When people organize wealth of perceptual information, they place it into categories that their language and culture give them. People somehow grasp some set of attributes that things have in common. On that basis they are grouped together into a category. Their language provides the symbol and conceptual categories that influence how its speakers' perceptions are encoded and stored. (Jandt, 1998, pp. 169–170, 175–176)

Some cultures may categorize certain things in great detail; others might not. The Eskimo culture, for example, has many categories for snow. (Jandt, 1998, pp. 185–186) In contrast, other cultures have fewer categories for snow. Another example is that in India an owl is a symbol of bad luck, the equivalent to a black cat, while the stork, which in the West is associated with birth, symbolizes maternal death in Singapore. It all depends on how people categorize them.

(3) *Interpretation*: This stage refers to attaching meaning to sense data and is synonymous with decoding. People give meaning to or "decode" the information that they have selected and organized. Diverse people can interpret the same information in completely different ways. (Jandt, 1998, pp. 177–178, 185–186)

Cultural differences in interpretation can be quite dramatic, as in the case of food. What one culture enjoys eating, another may find disgusting. Most people from Germany see dogs in the category of pet, and have learned that dogs are cute, cuddly, warm, loving creatures. When they see dogs as food in Korean, they will feel uncomfortable and disgusted. They think that dogs are pets, not food. Dogs are pets in some cultures and food in others. In the Arabian world, dogs are acceptable as watch dogs and as hunting dogs, but are not kept in the home as pets because they are seen as unclean and a low form of life. (Jandt, 1998, pp. 177–178)

Interpretation of colors also varies across cultures. White in the West is usually associated with weddings; whereas in China, Japan and India this color symbolizes mourning. Green, a color normally associated with freshness and good health in the West, is sometimes associated with disease elsewhere. Black is often seen as the universal color of mourning; however, in many Asian countries it is white, in Brazil purple, in Mexico yellow, and in parts of Africa dark red. Red suggests good fortune in China, but death in Turkey.

Although culture is very useful to human beings' perception in many different ways, it is also a problem in some respects (e.g., ethnocentrism, stereotypes, native realism), when people encounter something from other cultures.

(1) *Ethnocentrism*: Although cultures change slowly over time, a particular culture will tolerate a rather narrow set of behaviors and values. Other behaviors and values are labeled as wrong or foreign. This is what is known as ethnocentrism – people think that their own ways are superior, and that all other ways should be judged in terms of their ways. (Nolan, 1999, p. 13)

(2) *Stereotypes*: Since people learn about their culture's way of constructing the world when they are very young, cultural imprint is part of an individual's personality, just as culture created a "we" identity and a "they" category for all others. Stereotypes are summary generalizations about other culturally different groups. (Nolan, 1999, p. 13)

(3) *Naive realism*: Culture can give people the mistaken impression that they really know what is going on around them, because their culture is so much a part of who they are. They tend to develop an attitude of naive realism, the be-

lief that the way they see the world is the way the world really is. (Nolan, 1999, p. 13)

None of these problems helps people very much in their first encounter with a different culture. They are drawn, almost instinctively, to interpret what is happening in terms of their own cultural values and categories.

#### 4.2.1.2 High vs. Low Perception

A core question about perception of cultural difference is to what extent people can perceive it. The perception of cultural difference is a richer field than one that only discusses differences. Cultural differences are embedded within similarities, and cultural similarities are embedded within differences. Depending on how people look at it, similarities can look greater or the differences can look greater.

*High perception* indicates that the heterogeneity between cultures is perceived to be large, while *low perception* indicates that it is perceived to be small. Thus, cultural similarity/difference could be quantified, but they cannot be interpreted without thinking about either practical or theoretical importance.

Various effect size measures can be employed to look at how "big" a cultural difference is. For an example, Cohen (1988) used $d$ (ratio of difference in means to standard deviation) and $u_1$ (percentage of covered population) as two quantitative indices of difference between cultures. (Cohen, 1988. p. 21) According to the value of $d$ and $u_1$ the cultural difference is divided into three groups, small ($d < .2$, $u_1 < 15\%$), medium ($d < .5$, $u_1 < 33\%$), and large ($d < .8$, $u_1 < 47\%$). The medium level means that cultural difference may be conceived large enough to be visible to the naked eye. (Cohen, 1988. pp. 22, 26) However, $d$ and $u_1$ are indicators of the size of differences, not their acceptance.

Another example of empirical work to measure the size of cultural difference is Glaister's study. In this study, the measure of cultural similarity/difference was based on the perceptions of UK respondents and aimed at determining to what extent the national culture of foreign partners was different to that of the UK. (Glaister, 2004, p. 155)

### 4.2.2 Acceptance

#### 4.2.2.1 Acceptance of Cultural Difference

*Acceptance* means that people have a fundamental neutral stance towards cultural difference – seeing the perceptual realities of other cultures as valid. In other words, if people accept cultural difference, they will feel "the way that people do things is simply the way people do things." (Shaules, 2007, p. 166)

Acceptance implies an ability to *relativize* an experience and this often leads to a perceptual *decentering* as standards for judging a given cultural phenomenon shifts away from oneself and moves to larger frames of reference. (Shaules, 2007, p. 166) This means if people are open to accepting cultural difference, they should have internalized many other cultural norms into their own cultural frame of reference and enlarged it. Their own original culture is no longer the central set of standards for judging cultural phenomena. Instead of it, they conceive these new norms as normal and valid in the sense of being usual, rational, logical, understandable, appropriate, and preferable. (Torbiörn, 1988, p. 169)

Acceptance does not require that people change their own behavior to fit new standards. On the one hand, people could choose to change their behavior to fit new standards and experience the perceptual frameworks of others as fundamentally valid. In this case, other reasonable cultural standards are considered as a choice. On the other hand, they could choose not to change their behavior. In this case, it must not be because of a knee-jerk reaction to resistance. (Shaules, 2007, p. 167)

For some, the ideal of acceptance sounds suspiciously like a moral relativism in which "anything is okay". This reaction, while a misconception of acceptance, is understandable. In many studies, cultural choices are framed in terms of choosing one culture over another, and adopting someone else's cultural standards is the equivalent to losing one's cultural compass. Actually, acceptance involves increasing the number of conditions that must be kept in mind when adjudging cultural phenomena. Furthermore, acceptance involves an attempt to understand the cultural standards of others without assuming that they are inherently inferior. Indeed, acceptance implies that people cannot fully understand the cultural standards of others if they start from the assumption that they are inferior. (Shaules, 2007, p. 167)

Here people must make a distinction between difference that which they like and that which they accept, because it is easier to accept cultural difference as valid if it appeals to people personally. People could like some cultural differences without accepting them as a valid alternative for others. It is also possible to dislike some differences, yet still accept them. (Shaules, 2007, p. 168)

#### 4.2.2.2 High vs. Low Acceptance

The degree of acceptance can be roughly divided into two stages: first, *low acceptance with deep resistance* and, second, *high acceptance*. Many people seem to be able to accept cultural difference on the surface and do not have serious problems when encountering cultural difference. Yet, there is a big difference between those who stop at this level of acceptance and those who go deeper. (Shaules, 2007, p. 169)

(1) *Low acceptance with deep resistance*: People react to cultural difference only with low acceptance; in addition they are also engaged in a deep resistance. Low acceptance is often combined with deep resistance.

*Low acceptance* means that people accept certain cultural differences only to a low degree. With low acceptance people do not need a new cultural perspective to understand other cultures. Instead, some find a way to equate their own cultural system to other cultural systems they are familiar with. (Shaules, 2007, p. 173)

*Resistance* could be described as a long-term "bystander" phenomenon. This may be managed by isolating oneself from deeper adaptive demands. People are unwilling or unable to confront deeper adaptive demands from an insider's perspective. They only communicate with people who have the same cultural background as themselves. (Shaules, 2007, pp. 169-170) Resistance is considered *deep* if cultural differences that people are isolating themselves from are values and worldviews, which fundamentally created a culture and are considered as standards for adjudging cultural phenomena. (Shaules, 2007, p. 170)

Having a mixed reaction with low acceptance and deep resistance does not preclude, however, people needing intellectual knowledge in order to understand other cultures. For example, learning a culture starting with its history could give people lots of clues to current behaviors, customs and values, and trying to learn how people from other cultures think and behave. (Shaules, 2007, p. 171) However, an intellectual attempt to understand behavior is not equivalent to the ability to step into that worldview. (Shaules, 2007, p. 173)

(2) *High acceptance*: This requires not only the ability to see particular elements of cultural difference as valid, but also to accept the fundamental premise that other worldviews have a consistent internal logic. High acceptance implies seeing how implicit values and expectations work from the inside, not a rationalization based on external standards. This implies more than simply learning what to do in order to get the things one wants. (Shaules, 2007, p. 173)

By and large, people who managed to look at cultural difference with high acceptance are those that integrated themselves more fully into other cultures and who used a foreign language to do so. Since use of a foreign language implies an adaptation of the conceptual framework of a foreign culture, this seems understandable. (Shaules, 2007, p. 177)

In real life high acceptance of cultural difference is still rare, even among highly educated, experienced and successful interculturalists. This is because (1) people cannot accept cultural difference at a deep level until they have engaged with it. Many people who have extended experience abroad are not required to face deep adaptive demands. (2) If there are deep adaptive demands, people often have the ability to "pick and choose". That means enjoying the parts of their experience which they like and criticizing those that have found inconvenient or difficult to deal with.

### 4.2.3 Perception-Acceptance Matrix Model of Intercultural Adaptation

In an intercultural adaption process people perceive and accept foreign culture in different degrees, which range from high to low. Based on this knowledge the author developed a Perception-Acceptance Matrix Model to describe stages of intercultural adaptation. (See Figure 6)

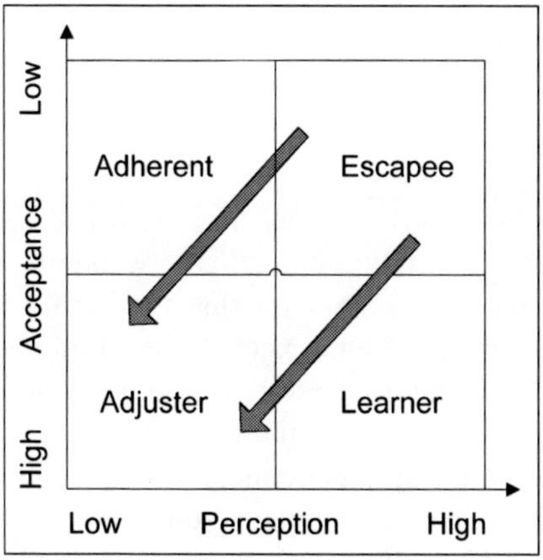


*Figure 6: Perception-Acceptance Matrix Model of Intercultural Adaptation*

Source: Own illustration.

*Perception* of cultural difference (horizontal axis) indicates the degree to which a person's own cultural frame is consistent with another cultural frame. In other words, people could notice some differences between their own culture and other cultures to certain degrees. In this regard, this model proposed that people tend to start at a high point of perception, which gradually decreases over time to a low point. In this process they acquire more and more intercultural knowledge and experience.

*Acceptance* of cultural difference (vertical axis) refers to the extent to which people accept other cultural frames; in other words, they see other cultures as valid. In this regard, this model proposed that people tend to start at a low degree of acceptance, which increases over time until they reach a high degree of acceptance after acquiring more and more knowledge and experience.

Combined perception and acceptance dimensions, the position of intercultural adaptation may vary within four areas:

- *Escapees*: high perception and low acceptance
- *Learners*: high perception but high acceptance
- *Adherents*: low perception but low acceptance
- *Adjusters*: low perception and high acceptance

People who adapt new cultures most successfully, referred to as "adjusters", must pass through three stages: ethnocentric (escapees), ethnomixed (learners/adherents), and ethnorelative (adjusters). And they pass along two developmental lines: escapees-learners-adjusters and escapees-adherents-adjusters. Short descriptions of each area of development are given below.

- **Stage I: Ethnocentric stage**

*(1) Escapees*

The starting point of two developmental lines is escapees. In this stage, people avoid confronting a foreign culture and remain outsiders. Cultural difference is large, obvious and sometimes even surprising for them, because they live relatively isolated from other cultures and know little about other cultures. They still cannot accept them, but begin to be aware of the fact that other cultures are different compared to their own. People in this stage use the standards and customs of their own culture to judge all other cultures unconsciously and describe other cultures using stereotypes based on knowing only one or two things about them. (Bennett, 2007, p. 26) For example, many U.S. Americans seem to think that all Africans live near jungles and have encounters with wild animals; or many Asians seem to think that all Americans from the Pacific Northwest live on ranches and ride horses.

- **Stage II: Ethnomixed stage**

The middle stage of the two developmental lines consists of *learners* and *adherents*. In this stage, people have moved beyond ethnocentric rules and have not yet developed ethnorelative principles. They accept primarily their own cultural frames of reference, but they begin to perceive and accept one or more different cultural frames. Most people at this stage are generally interculturally sensitive with varying degrees of perception and acceptance.

*(2) Learners*

Learners consider cultural difference to be large, but they keep an open mind and spend time learning it and try to accept it. They recognize and accept superficial cultural differences such as food culture and other customs, but they assume that deep down cultural difference is essentially still very large. They learn to incorporate selected elements from another cultural system into their own personal cultural frames. Even for people who are highly committed to cross-cultural learning, this process can be slow and difficult. (Nolan, 1999, p. 25)

People in this stage enjoy recognizing and exploring cultural differences. They are aware that they themselves are cultural beings. They are fairly tolerant of ambiguity and are comfortable knowing there is no unique right answer. Acceptance does not mean that a person has to agree with or take on a cultural perspective other than his or her own. Rather, people accept different cultural ways of thinking and behaving, even though they might not like them. This is the first stage in which people begin to think about cultural relativity. This means their own cultural frames are not the only good way to be in the world. In this stage people may become paralyzed by the dilemmas posed by conflicting cultural norms.

*(3) Adherents*

Adherents, on the other hand, consider cultural difference to be small, but they do not accept it. People use knowledge about their own and other cultures to intentionally shift into a different cultural frame of reference. That is, they can take another cultural perspective in order to understand things across cultural boundaries based on their ability to use alternative cultural interpretations. They try to bury cultural differences within already-familiar categories and thus cultural difference becomes smaller and smaller for them.

Although they start to be able to understand another culture successfully based on more learning and they developed more ability to construe cultural difference, in this stage people are still not prepared to change their stable worldviews and values. They view their own culture as "center" and tend to evaluate different cultures as "marginal".

- **Stage III: Ethnorelative stage**

*(4) Adjusters*

The end point of the two developmental lines is the adjusters. Finally, at this stage, people cannot clearly recognize cultural difference and they can totally accept cultural difference. They tend to convergence their own culture and other cultures in their minds. People have internalized one or more cultural frames in addition to that in which they were originally socialized. As a result, they adopt a new more integrative culture which combines several different cultures and embrace some aspects of new cultures. Little by little, they become comfortable with many standards and customs, and have an ability to adapt behavior and judgments to a variety of cultures.

As they move into the adjust stage, they see themselves as interculturalists in addition to their national and ethnic backgrounds. They become overwhelmed by the cultures they know and can completely shift their cultural frame of reference without much conscious effort. Thus, they are inclined to interpret and evaluate cultural phenomena from a variety of cultural frames of references, so that there is never a single right or wrong answer.

## 4.3 Cultural Knowledge and Intercultural Adaptation

To shift from the ethnocentric (escapee) to ethnorelative (adjuster) stage, people need accumulated knowledge about other cultures based on an intercultural mindset. The recognition that something is culturally different requires an understanding that there are different cultural systems at play. In order to understand differences, people must not only have the ability to set aside their own judgments, but also long-term experience and knowledge about a new way of thinking and doing things. (Shaules, 2007, p. 174)

A cross-cultural mindset helps people look behind facts and figures to uncover meanings and patterns, learn in unfamiliar cultural phenomena, and gain entrance into the cultural world of others. Without this ability, people observe another culture only as outsiders and are always subject to someone else's understandings and interpretations. (Nolan, 1999, p. 2) Having this mindset as part of their personal inventory will then enable people to understand and accept cultural difference.

People who have a cross-cultural mindset generally recognize that all cultures have coherence and an implicit rationale, and no one culture is inherently better or worse than another. (Nolan, 1999, p. 31) They share a set of traits or attitudes, which include the following:

(1) *Flexibility and adaptability*: A willingness to bend, to try new things, and to change routines and patterns as a result of learning.

(2) *Tolerance and patience*: Tolerance not just for difference, but for ambiguity and mixed signals that are so much a part of the start of a cross-cultural experience. This is coupled with patience to let meanings emerge slowly as experiences accumulate.

(3) *Intellectual and social curiosity*: A genuine desire to know more about people and their way of life, their ideals, and their value and costumes. (Nolan, 1999, p. 31)

Based on a cross-cultural mindset people accumulate the knowledge about other cultures little by little. Several studies revealed that knowledge about other cultures is related to the degree of intercultural adaptation. Ward found in his empirical study that people's knowledge about other cultures and languages is one of the key factors facilitating their cultural adaptation in other cultural environments. (Ward, 1999) In Black's study, the findings revealed that pre-departure cultural knowledge about the host country is related to people's intercultural adaption in a new cultural environment in the host country. (Black, 1988)

There are basically two types of knowledge that gives people the tools to understand and accept cultural difference: *content* – the specific details of costumes, values, way of life, behavior patterns and so on – and *context* – an understanding of environment that surrounds the activity to be carried out. Understanding context enables someone to know why this behavior is appropriate for this particular situation. People who possess content knowledge will always be able to understand cultural difference superficially, while people who possess context knowledge will be better able to accept cultural difference.

# 5 Summary of Part B

## 5.1 Summary of Theoretical Framework

This part of the theoretical framework consists of three aspects: cultural difference; cultural difference in TV programs; and perception and acceptance of cultural difference.

*(1) Cultural difference (Chapter 2)*

Culture is that complex whole which includes knowledge, beliefs, art, morals, customs and habits which are acquired as a member of particular society. Cultural differences vary in quite different ways which are divided according to horizontal dimensions and vertical levels.

Each *horizontal cultural dimension* has two poles to describe opposite situations and a particular culture position somewhere within the range between two poles. Several famous theories about cultural difference are introduced: Hofstede's Five Cultural-difference dimensions, Kluckhohn's Basis Cultural Orientation Dimensions, Parsons' Cultural Pattern Variables, and Hall's High- and Low-Context Culture. *Vertical cultural levels* consider the hierarchical relationships between cultural dimensions in an iceberg model and an onion model.

*(2) Cultural difference in TV programs (Chapter 3)*

Television as an everyday medium demands more or less proximity with the life world of viewers. Fundamental areas of television programs, such as character, plot construction, dominant themes, ideology, language, and humor may have totally different meanings resulting from cultural interpretation.

There are two controversial theories about cultural difference in mass media: cultural discount theory and universal media culture theory:

*Cultural discount*: A particular television program rooted in one culture, and thus attractive in the home market where viewers share a common knowledge and way of life, will have a diminished appeal elsewhere, as viewers will find it difficult to identify some unfamiliar or unexperienced culture in foreign programs. Cultural discount varies across television genres and also depends on the cultural proximity between the country of origin and the country of destination.

*Universal media culture*: To appeal to global audiences comprised of various linguistic and ethnic groups, television programs require cultural-neutral attributes with a universal appeal to maximize audience size and revenue. They could improve mutual knowledge among a global audience and increase their consciousness of life they have in common and develop their sense of belonging to a world universal culture community. A result is cultural imperialism, which is frequently used to sum up the domination of American media products.

*(3) Perception and acceptance of cultural difference (Chapter 4)*

Intercultural adaptation is the entirety of the phenomena of individuals, who strive to adjust their mind and behavior to an unfamiliar cultural environment, establish and maintain a relatively stable and functional relationship with a new culture. The goal of intercultural adaptation is achieving a "fit" between one's own culture and an unfamiliar new one. Some theoretical models about cultural adaptation include Lysgaard's U-Curve Model, Adler's Five Stages of Transitional Experience, and Bennett's Developmental Model of Intercultural Sensitivity.

All of these models are concerned with to what extent people can perceive and accept cultural differences. Perception and acceptance are two different concepts for describing people's feelings and opinions about cultural difference. Perception indicates the size of cultural differences, whereas acceptance indicates the degree to which people regard other cultural frames as valid.

Combined high/low perception and high/low acceptance the author developed a Matrix Model to describe the stages of intercultural adaptation. The intercultural adaptation process passes through three stages: ethnocentric (escapees), ethnomixed (learners/adherents), and ethnorelative (adjusters). To shift from ethnocentric to ethnorelative stage people need intercultural mindset to accumulate knowledge about foreign cultures.

## 5.2 Research Limitations

### 5.2.1 Limitation of the Cultural-Difference Dimension

*(1) Cultural-difference dimension for media study*

The various concepts of cultural-difference dimensions, such as individualism, time orientation, relationship with nature, and innate human nature, have some limitations if they are used in media studies, because these dimensions are only concerned with living environment and are not directly related to media products. Although these cultural differences are indirectly reflected in television programs through drama plots, dialog and behavior of characters, narrative styles and so on, they are too abstract and difficult to understand for media users. Thus, when conducting media studies on cultural difference in media products, people need a set of understandable dimensions, which are interpreted in a way that is familiar to media users and which they can understand.

*(2) Overview of cultural-difference dimensions*

In many media studies cultural difference was often regarded as a whole. Various types of cultural difference were never summarized systematically and never done within a comparative framework. Few studies provided an overview of the various types of cultural difference for different television genres. Actually, the case for fiction may not necessarily be the case for news. Some studies only tested a separate subset of genres. For example, they only paid attention to fiction genres, rarely non-fiction genres. The more cultural-difference dimensions there are involved, the more reliable and valid are the explanations about a fuller account of culture difference.

What this study advocates is that scholars should find a way to navigate between extreme overgeneralization and extreme narrowness without being too broad or too restricted. The analysis of cultural difference in media products should be differentiated and people should take a rather comparative view in regard to the various types of differences. Without this broader comparison the discussion of cultural difference seems to be rather limited.

*(3) Cultural background of cultural-difference dimensions*

The two most popular empirical designs for examining intercultural adaptation, quantitative approach and qualitative approach, do not serve very well as a way to examine cultural difference – each one for a different reason.

Many studies about the reception of television programs relied on quantitative dates of commercial audience research such as viewing ratios, which are collected using modern statistical techniques. However, they could only indicate the results of the diminished consumption of foreign television programs due to cultural difference; they cannot supply the information about the cultural reasons lying behind diminished consumption; in other words, why cultural difference causes diminished consumption.

Besides quantitative audience research, the new electronic media technology offers a substitute and a source of feedback that goes beyond ratings information. Television viewers can directly respond to programs through chat-rooms or websites dedicated to them. The opinions, tastes, and trends of television viewers could be analyzed based on this feedback. However, this qualitative approach is rarely feasible because feedback gathered this way is often short and not detailed enough for it to be used to analyze the cultural background of the viewer. Furthermore, samples are random and not representative, which means that this approach is not as reliable as quantitative rating research. Usually researchers face enormous difficulties in attempting to measure and control statistically for all relevant variables that might make different people roughly comparable.

*(4) Quantitative measure of cultural-difference dimensions*

Many studies used the *Kogut Singh Index* to quantitatively measure the degree of cultural difference between nationalities, in which five cultural dimensions defined by Hofstede were aggregated. (Glaister, 2004, p. 155) However, in this measure there is no guide in an absolute sense as to what is near or what is distant – there is only a relative distance measure based on a comparison with other nationalities. In other words, the cut-off point for large and small cultural difference is not defined.

For example, in the study by Glaister (2004) three nationalities – Italian (.91), Dutch (1.87), and French (2.27) – are regarded as the *nearest* cultures group. These three nationalities with the lowest index were chosen arbitrarily as the *most similar* cultures. Among the remaining partner nationalities, the *most different* cultures group ranged from German (2.34) to Swedish (2.50). However, how large the index of 2.50 actually is, could not be defined. (Glaister, 2004, p. 155)

*(5) America as research center*

In many studies the culture of the United States is often considered as the center. Usually they have a home base country – the United States (Here) – and they are interested in communication with other countries (There). (Chaffee, 1992, p. 37) Though it is generally true that U.S. media exports tend to dominate world markets, it is also valuable to know how audiences respond to media products from other countries. For example, French and German films are more favored in certain societies than others.

### 5.2.2 Limitation of Perception and Acceptance of Cultural Difference

*(1) Evidence of reception*

Cultural discount theory indicates that a particular television program rooted in one culture will have a diminished appeal elsewhere due to various cultural differences. Universal media culture theory argues that international television programs require cultural-neutral attributes with universal appeal in order to maximize audience size.

Although both theories are plausible at face value and there are even a handful of empirical studies on this issue, they provide scant evidence about the reception of cultural difference or universal media culture; in other words, to what extent media users could accept it. One reason is that it is very difficult to identify the components of cultural difference to examine their acceptance.

*(2) Evidence at individual level*

Examination of cultural difference is often based on country level, for example, movie box-office statistics or television-viewing ratios of particular countries. However, few studies empirically addressed cultural difference from an audience perspective at an individual level. A review of the literature shows that culture difference in foreign media at individual level has not been subjected to empirical testing.

When dealing with cultural discount or universal media culture, people need to acknowledge that the use of media products occurs at the individual level, and different media users perceive the same product differently. Therefore, the best way to approach cultural difference in media products may be to analyze it at an individual level. This means that studies should pay close attention to opinions and interpretations of individual media within their own cultural framework.

*(3) Translation of attitude into behavior*

Although there are many studies concerned with the sequence and forms of intercultural adaptation in cross-cultural situations, these models do not seek to explain how the worldview of "adaptation" translates into actual adaptive behavior. (Landis, 2004, p. 251)

Interculturalists believe that in a social-science context action tends to be associated with attitude – a more or less positive or negative feeling about an object of perception and, possibly, a predisposition to action. (Landis, 2004, p. 251) Understanding cognitive constructs such as the values of a target culture can be translated into actions, such as modified social behavior in that culture.

# Part C: Design of Empirical Study

# 6 Television in China

In this study China is used as a sample country for researching cultural differences in universal television programs for several reasons:

(1) Chinese culture is a typical high-context culture with high collectivism, whereas the culture of many Western developed countries (e.g., U.S. and Europe) is low context with high individualism. Thus, the contrast between Chinese culture and culture in universal television programs from Western developed countries might be very obvious and interesting.

(2) Since *Reform and Opening-Up Policy* in 1978, China's society has changed radically and become more open to the outside world. Little by little, people in China have changed their ways of life, habits, values, worldviews, and so on, partially according to Western standards. China now stands at a "crossroads" of economic, social and cultural change.

(3) In the last thirty years the Chinese broadcasting system has transformed from "single track" with a *propaganda* function to "double track" with not only *propaganda*, but also *commerce* functions. This means that Chinese television stations must compete with each other to survive in the market. Thus, they must broadcast attractive but relatively cheap universal television programs from developed countries and Chinese viewers can watch more and more foreign programs.

(4) The author knows both high-context and low-context cultures very well, because she grew up in China and studied in Germany for nine years. Her native language is Chinese, so she does not have communication problems with Chinese participants.

## 6.1 Transition of China's Society

Since the *Reform and Opening-Up Policy* in 1978, Chinese society has changed radically. The change from a *Seclusion Policy* (BiGuan SuoGuo) to an *Opening-Up Policy* (GaiGe KaiFang) makes various social areas in China more open to the outside world. The switch from a *Planned Economy* (JiHua JingJi) to a *Socialist Market Economy* (SheHui ZhuYi ShiChang JingJi) promotes competition in economic areas in China. The *Confucianism*-dominated Chinese traditional culture has gradually become integrated with global culture.

*(1) Social change*

The *Seclusion Policy* dominated in China for several centuries. The main purpose of this policy, which cut off China from the outside world, was to keep China's powerful neighbors out and remain aloof from any foreign complications, such as unorthodox ideals. (Cheong, 1997, pp. 6-10)

*(2) Economic change*

After the founding of the People's Republic of China in 1949, the question arose of how to feed such a huge population. To tackle this *Mao Zedong* launched the *Planned Economy* (JiHua JingJi), which was similar to the Soviet model. In the planned-economy system the organizational structure, production, revenue and expenditure plans of a state's enterprises were decided upon and controlled by the government and some public institutions. Thus, the state's enterprises worked very inefficiently and could not satisfy the huge consumption needs in China. (Staiger, 2003, p. 477)

A consequence of this was that China's economy suffered from slow growth and stagnation. It faced challenges of capital shortages, out-of-date technology skills, foreign-exchange restrictions, and employment pressure. With the new science and technological revolution driving the world economy at a faster pace, China found itself lagging farther behind in economic as well as scientific and technological strength. (Zeng, 2010)

The *Reform and Opening-Up Policy* was initiated in 1978 by reformists within the *Communist Party of China* led by *Deng Xiaoping*. It refers to the program of social economic reforms aimed at building *Socialism with Chinese characteristics*. In the Third Plenary Session of the *11th Central Committee* held in December 1978, the *Communist Party of China* shifted its focus to socialist modernization and opening-up. In July 1979, China began experimenting in some special zones in *Shenzhen*, *Zhuhai*, *Shantou*, and *Xiamen*. The success of these four special economic zones provided valuable experience for China's economic development at that time. (Zeng, 2010)

In February 1984, the *Communist Party of China* decided that in addition to the existing special economic zones, more *Economic and Technological Development Zones* (JingJi JiShu KaiFa Qu) in coastal areas in China should be opened up to the outside world. The main purpose of this was to take advantage of their geographic locations, build up a favorable business environment to attract foreign investment to these small special regions as soon as possible and promote an export-oriented economy. (Zeng, 2010)

Since the launch of the *Reform and Opening-Up Policy* the Chinese government gradually realized that market and competition are the essential factors to advance economic development in China. In 1980, for the first time the State Council issued a policy to protect and promote competition in the market. Since

1984 the prices and quantities of consumer and production goods have been determined not by the government alone, but also increasingly by the market. Thereby, the transition from a *Planned Economy* to a *Socialist Market Economy with Chinese characteristics* (ZhongGuo TeSe SheHui ZhuYi ShiChang JingJi) became anchored in the Constitution in 1993. Thereafter, a mixed form of state and private ownership of Chinese enterprises increasingly emerged. (Staiger, 2003, p. 477)

*(3) Cultural change*

In the past thirty years, there have not only been social and economic changes, but also cultural changes in China, such as changes in values, worldviews, ways of life, habits, and customs. The traditional Chinese culture is currently facing a cultural challenge from the outside world. China is now a land of sharp cultural contrasts and contradictions. As a result of globalization something traditional is being lost in the rush to modernize and develop.

In traditional Chinese culture the most important social values are derived from Confucianism, which was the official philosophy throughout most of imperial China's history. Confucianism focuses on cultivation of virtue and maintenance of ethics, the most basic of which are *Ren* (Humanity), *Yi* (Righteousness), and *Li* (Propriety or Etiquette). (Bell, 2008, p. 195) *Ren* is an obligation of altruism and humaneness for other individuals within a community. *Yi* is upholding righteousness and moral disposition to do something good. *Li* is a system of norms and propriety that determines how a person should act appropriately within a community. (Shen, 2007, pp. 8-11) Besides Ren, Yi, Li, there are many other cultural virtues in Confucianism, such as *Zhi* (Knowledge), *Xin* (Integrity), *Zhong* (Loyalty), *Xiao* (Filial piety), *Jie* (Continency). (Mou, 1999, p. 103) Confucianism is humanistic and non-theistic. So it does not involve a belief in the supernatural or in a personal god.

In practice, Chinese examine the world using the logic of humanity, righteousness, and propriety as an ethical philosophy and behave according to this framework. The Chinese traditional core values and worldviews, such as harmony, collectivism, family-centered values, and respect for seniors are fundamentally derived from these Confucian thoughts. In China people learn traditional culture from childhood from books, teachers, parents, relatives, and friends.

However, during the period of globalization in the last thirty years, Chinese traditional culture has become lost little by little through radical development and opening up to the outside world. Yet today, *Coca-Cola*, *McDonalds*, and S*tarbucks* dot the street corners of large cities. Eating foreign fast food, watching foreign movies, listening to foreign music, and wearing clothing with foreign brands has become the modern lifestyle for the young generation in China. They either do not have many opportunities to learn Chinese traditional culture, or they forget it gradually, because their life is now fulfilled by foreign products and cultures. Many think that foreign cultures are very attractive, whereas Chinese traditional culture is boring and belongs to their parents and grandparents. Young people in China may have similar habits and ways of life to young people in America or Japan, but most do not have or have lost their traditional Chinese cultural identity.

## 6.2 Transition of China's Television System

The construction of a television system in the People's Republic of China began in the mid-1950s. In May 1958 the first television station in China *Beijing Television Station* started to broadcast experimental programs. The television network gradually began to expand in the early 60s. Ten years later in 1973 saw the launch of color television. Since the mid-1980s every television station in China has been able to broadcast in color. (Staiger, 2003, p. 216)

Since the late 80s television has played a central role in political communication in China. Based on its large broadcasting scope (90% of the population in China) television has become the dominant medium above all others. (Staiger, 2003, p. 206) As a relatively young medium, television has quickly developed and permeated into China, and gained political, economic and cultural significance in society. With over one billion television viewers (range: 93.6% of population), more than 90 million cable-TV subscribers and 350 million television sets, the People's Republic of China is now the world's largest television market. (SARFT, 2002, p. 45)

### 6.2.1 Double-Track Television System

During the period of the planned economy in China, television was completely financed by the state. Its monopolistic management model was characterized through the strict revenue and expenditure plans and lack of commercial motivation. All television programs were produced and provided by the state. At that time, television had only one function – a propaganda instrument of the *Communist Party of China*. Program content was limited to transmitting propaganda that advocated party and political interests.

The economic reforms after the *Reform and Opening-Up Policy* in 1978 also had radical consequences for Chinese television. Television commercialization is also a part of the development of a market economy, because "reform and openness since the 1980s had created a growing demand by foreign and domestic enterprises for effective advertising channels". (Zhao, 1998, p. 52)

With the broadcast of the first TV advert in Shanghai in 1979, television in China started to change – but without losing its political status. The *Beijing TV Station* was renamed *China Central Television* (CCTV). This station had a dual role: On the one hand, it was still a state institution and a propaganda instrument to serve political needs; on the other hand, it is quasi-enterprise which had to be run more efficiently and was permitted to generate income from the market to a limited degree. (Ollig, 2002, p. 9) Later, the *double-track television system* was launched in China, whose content not only has a political function, but also economic.

In this new system, Chinese television stations are so-called *ShiYe DanWei* (public institutions) with *QiYe GuanLi* (entrepreneurial management) (Fischer, 2001, p. 3). The term *ShiYe DanWei* refers to public institutions that focus on the production and provision of public goods and services. They receive subsidies from the state and receive preferential treatment regarding taxation. As public institutions, television stations are not required to be entered into the registry for businesses and do not have to pay sales tax. The term *QiYe* refers to enterprises that produce private goods and services. They can generate profits through the market and do not receive state subsides. The slogan *ShiYe DanWei, QiYe GuanLi* means that television broadcasters are public institutions by nature, but they are run as quasi-enterprises. (Ollig, 2002, p. 31)

Chinese television simultaneously operates upon two different modes of logics: the party's and the market's. (Zhao, 1998, p. 84) Television stations must, on the one hand, fulfill propaganda tasks since they are state property and, on the other, pursue economic goals. However, one of the two modes of logic holds priority over the other. In the hierarchy of targets, propaganda goals always have top priority. Their vision and mission are primarily defined by constitutional standards and state contracts based on political interests. Economic goals are always secondary. The reason is that "the political monopoly of China's Communist Party has not been successfully challenged, and information control is still one of the remaining instruments of the Communist Party to sustain its political power." (Fischer, 2003, p. 5)

### 6.2.2 Propaganda Function

China has a tradition of propaganda, which stretches back more than 2500 years. The Chinese government has always used propaganda throughout its long history to promote what it believed to be true and correct. Behind this hid the Chinese traditional way of thinking and behavior – propaganda as an educational approach was considered more effective than use of force or punishment. This approach is based on a Confucian belief that people tend to imitate moral paragons. Thus, from the beginning the most important task of Chinese television was to educate the audience according to the government's interests.

The propaganda function of Chinese television is implemented through strict regulations by the government. There is no program free from censorship. The content of television programs must correspond exactly to the policies and goals of the government. Censorship guaranteed that inappropriate programs, which would have a negative effect on the government and society, cannot be broadcast.

*(1) Institutes of censorship*

Strict censorship is implemented by the *four-level administrative system* at the center, province, city, and circle level:

At the central level the *Propaganda Department* (XuanChuan Bu) of the Communist Party is responsible for television censorship. It ensures that TV stations are under the control of the party and act according to political policy. This department is supposed to stabilize political actions and ensure that only suitable information from the outside world (e.g., Western countries) penetrates into China. (Fischer, 2001, p. 3)

Alongside the *Propaganda Department* the *State Administration of Radio, Film and Television* (SARFT) as the highest state authority to administer radio, film and television has the most significant impact on Chinese television at the central level. Its main task is day-to-day censorship work in approving and supervising all kinds of program content. (SARFT, 2006)

*(2) Guideline of censorship*

Censorship is based on a lot of guidelines in which permitted and prohibited content is defined. Permitted content includes, for example, television programs which serve and are useful to society, distribute scientific and cultural knowledge to increase the educational levels of citizens, are good for social and economic development, and enrich people's spiritual lives. (Fischer, 2001, p. 3) Prohibited content includes, e.g., anything dealing with superstitions, pornography, violence, anything contrary to the principles of the constitution and party, anything that threatens the security and reputation of the state. (Fischer, 2001, p. 3)

Fundamentally, censorship of Chinese television runs strictly according to the *Four Cardinal Principles* (SiXiang JiBeng YuanZe) (Ollig, 2002, p. 25), which were stated by *Deng Xiaoping* in 1979. These are: upholding the socialist path; upholding the people's democratic dictatorship; upholding the leadership of the Communist Party of China, and upholding Marxist-Leninist and Mao Zedong thought. (Li, 1995, p. 438)

*(3) Censorship for different television genres*

Censorship varies across television program genres. News programs are the most strictly controlled genre, because they represent the most important propaganda tool for government. Drama has more freedom than news programs, but even in this entertainment area there are still some strict controls.

*News programs*: At the central level, the *Propaganda Department* of the Central Committee of the party has the highest authority to supervise news programs. In news programs, the party primarily advocates topics, which promise to generate a positive image for the party and legitimize its political decisions and measures. (Jarren, 2002, p. 111) The control of national news and international news lies in the hands of the state news agency *Xinhua* (New China), which is headquartered in Beijing and has a monopoly of news distribution in China. In the editorial department of *Xinhua* there is a party secretariat, who supervises the output to ensure the party's guidelines are being adhered to in news programs.

At province, city, and circle levels, the journalists, television stations, and propaganda departments of local government must also implement self-censorship. Through education, personal experience and internal censorship journalists must be able to determine what is considered to correct from the perspective of the party. (Ollig, 2002, p. 36) The former director of the *News Commentary Department* of *China Central Television* Sun Yusheng stressed: "During the process of previewing the programs, I keep thinking about the following questions: Will these produce negative effects? Will it cause damage to political stability? Will it intensify tensions and be detrimental to solving the problem? After all these possibilities are eliminated, I will say that an item is a good one and should be broadcast." (Eickhoff, 2004, p. 82)

*Drama*: Although drama has more freedom than news programs, it must also be supervised. The Chinese government emphasizes that television drama is primarily there to distribute political, economic and cultural guidelines which are issued by the party's Central Committee and the central government. (Donald, 2002, p. 36)

The whole produce process for a drama is strictly controlled. Drama producers must first apply for a license from the SARFT. Before the program can actually go into production the themes of the drama must be approved by the *Chinese Television Arts Committee*. During production the producer and the local party propaganda office must check by themselves if there is any inappropriate content. Finally, an already broadcast drama could be stopped if it fails to be sanctioned by the SARFT or the TV broadcasters. (Eickhoff, 2004, p. 82) In addition, the influence of television dramas from the U.S., Japan, Korean, and

Hong Kong has slowly brought about a transformation of drama from a pure political propaganda tool to an entertainment medium in China.

*(4) Censorship for foreign programs*

Foreign programs are strictly controlled in China. According to the SARFT guidelines foreign programs must fall in line with the requirements of the party's *Propaganda Department*. "The Chinese people are not seeing programs that offend Chinese sensibilities or challenge the party's worldview." (SARFT, 2006) To reduce influence from abroad the SARFT established a maximum import quota and supervises the content of foreign programs before they are broadcast. Particularly strong is the government's intervention regarding news imports. Foreign dramas are prohibited from being broadcast during prime time. All program imports must be shown by the *China International Television Corporation*, which subordinates *China Central Television*. Foreign programs cannot be broadcast without authorization. (Eickhoff, 2004, p. 36)

### 6.2.3 Commerce Function

Commercialization expresses the introduction of commerce in the Chinese television system. It vaguely describes that "some commodity or service that had not been traded for money in the past is now being traded for money." (Fischer, 2003, p. 7) It also describes a process of target shifting of Chinese television, which is no longer primarily focused on its content or its assumed social functions, but on the economic returns from its output. (Fischer, 2003, p. 7)

In practice the Chinese government acknowledges that television is not only the mouthpiece of the party, but that it can also achieve profits and survive in the market on its own. Since the withdrawal of government from the full financing of television in the early 90s, Chinese television stations have been in competition with one another and have gradually turned from being propaganda tools to quasi-enterprises. Commercialization firstly took place in certain lucrative program categories including dramas, sports programs, entertainment shows, and music programs. Other genres such as news, which are connected with propaganda, are still produced by the state.

To commercialize Chinese television the Chinese government switched from a protection policy to a competition policy. It has reduced its operational activities in regard to Chinese television; today its intervention is focused only on political and strategic guidelines and target setting. Thus, the central state regulation is changing to a partially decentralized self-regulation. It is expected that through commercialization Chinese television will run increasingly autonomously and independent of political actors. There are two approaches to commercializing Chinese television:

*(1) Reduction of the state's subvention and self-financing through the market*

The self-financing through the market firstly occurred in the 90s. At that time, Chinese television stations had a mixed financing form of state subvention and market financing. The government has since transferred financial responsibility for television to the market as far as possible. Nowadays, advertising revenues are the main source of income for Chinese TV stations. Only a few important TV channels and TV stations, especially in remote regions, get financial support from the state.

*(2) Competition policy*

The launch of the competition policy has had many positive consequences for the Chinese television market. It has optimized the whole television system in economic terms and enhanced competition between TV stations. Competition creates greater freedom for television stations to act better according to market logic. They can freely experiment with new market approaches and choose their own strategies, which previously were determined by the government.

Self-financing through the market and the competition policy require Chinese television stations to behave more economically. (Liu, 2001, p. 23) Both approaches led them to pay more attention to the audience's demands and to constantly optimize their programs in order to enlarge audience share and generate more advertising revenue. (Guo, 2004, p. 19) A simple and efficient way to attract a higher audience is to broadcast high-quality foreign television programs, which have been successful in foreign markets. This will be discussed in the following chapter.

## 6.3 Foreign Television Program in China

### 6.3.1 Market Opening

Effective competition in the Chinese television market is only possible if the opportunity exists for newcomers to enter the market. In the last thirty years the Chinese television market has been gradually opened up to foreign media companies. Market entry has become much easier compared to what it was under the planned economy based on relaxed regulations in this sector. The most important milestone in the opening up of the television market was China's accession to the *World Trade Organization* in 2001 which hold long-term significance for the development of China's television market. (Guo, 2004, p. 21) The partially opened market has only slightly improved the function of competition, yet the improvement is obvious.

*(1) Market opening and accession*

Already before China's WTO accession many multinational media companies (e.g., Murdoch's *News Corporation*) began to recognize the importance of China as a very big target market for their media products simply by absolute numbers. The potential television advertising market has been estimated at 25 billion US$, and this offers huge potential for foreign companies in terms of market share. (Bensinger, 2005, p. 7) Thus, they started negotiating with Chinese media companies on joint ventures decades before any regulation was actually adopted.

Chinese television stations, which now must be self-financed, are making an effort to corporate with foreign media corporations to produce high-quality programs to attract a higher Chinese audience and achieve greater advertising revenue. For example, they have adapted successful foreign television formats.

Since the WTO agreement the tendency is clear that the Chinese television market will be much more open. The entertainment program market was opened up first. Then, the core area of propaganda was also opened up increasingly to foreign suppliers and investors. (Lewin, 2004, p. 2)

In October 2001, the SARFT issued the first broadcasting rights to a non-Chinese television institution *Murdoch's Starry Sky Satellite TV* in China; however, the coverage area was limited to Guangdong province. The first Chinese programs produced by *Star TV* hit the air during the first half of 2002. *America Online* simultaneously launched its Chinese-language entertainment channel *China Entertainment Television*, which provides a mixture of films, sports and television series, but without news. By 2002, 27 foreign satellite channels already had broadcasting rights in China in hotels, in residential areas for foreigners, and selected academic institutions. (Eickhoff, 2004, p. 53) *CNN* and *BBC World* also finally received permission to enter the Chinese market, much earlier than expected. (Guo, 2004, p. 21)

*(2) Market limitation*

Market opening is not without limitations and control. The Chinese government allows market opening where it has been "identified as economically necessary" and as beneficial to securing its own power. (Ollig, 2002, p. 50) Although the market access barriers have been progressively eliminated, the market accession for foreign companies is at present very difficult and the degree of market opening is still not sufficient for effective competition between Chinese and foreign companies.

The market access of foreign programs is always controlled by the Chinese government and broadcasting permission must be given only under special consideration of the party's policies, because it recognizes the importance of foreign television programming as a supplement to the country's culture and society. (Ollig, 2002, p. 50) Thus, the Chinese government clearly emphasized in the *GATS* (General Agreement on Trade in Services) agreement that the operation of television stations will continue to be in the hands of the Chinese state in order to protect Chinese television stations from foreign competition and influence. (Eickhoff, 2004, p. 46)

The *Propaganda Department* of the Communist Party of China and the SARFT have the final say in approving program imports and international cooperation. This means that no broadcaster can access the Chinese television market without a permit from these state administration institutions. The political control is always a barrier for foreign newcomers accessing the Chinese television market.

### 6.3.2 Forms of Market Accession

The forms of market accession are diverse, e.g., direct importing, localization, co-production, and joint venture:

*(1) Direct importing*

*Direct importing* means that foreign television programs are directly imported to China and broadcast with a few modifications, such as subtitling or dubbing. All program imports must be handled by the *China International Television Corporation*. The SARFT supervises their content and decides on broadcasting quotas as well as an overall import quota of foreign programs. (Eickhoff, 2004, p. 36) Because of various cultural differences in foreign programs, direct importing is generally only possible for movies, television series, sports programs, and documentaries. Genres such as foreign news programs, talk shows, and entertainment shows are not usually accepted by Chinese audiences.

The major exporters to China are the United States, Japan, and South Korea. The *China TV Program Agency* as well as the *International Relations Department* as a part of CCTV imports a high proportion of television series and cartoons from Southeast Asian countries (Ollig, 2002, p. 50), while the U.S. and Europe have a greater share in terms movies, sports programs, and documentaries.

*(2) Localization*

*Localization* of foreign programs represents a more useful measure for Chinese television stations than importing finished programs, because through localization the foreign TV format can be "filled" with culturally specific content according to the preferences and habits of the Chinese audience by local staff.

(Guo, 2004, p. 21) Thus, the problem of cultural barriers between the Chinese audience and foreign producers could be better solved.

*(3) Co-Production*

In December 2002 Murdoch's *News Corporation* signed an agreement with the *Hunan Radio, Film and Television Group*, in which they agreed to co-produce a program. This agreement is important because it allowed the first co-production between a Chinese and a foreign media cooperation. (Bensinger, 2005, 12) SARFT's vice-president Zhao explained that co-productions mean a win-win situation for both sides. On the one hand, foreign media cooperation could understand the Chinese audience better; on the other hand, Chinese television could gain valuable experience and economic efficiency could noticeably increase through a "learning by doing" process. (Ollig, 2002, p. 50)

*(4) Joint venture*

In 2003 the first joint venture between *Viacom/Nickelodeon* and the *Shanghai Film Group* was permitted to broadcast a children's television program through a cable network in China. (Bensinger, 2005, 12) Through joint ventures, Chinese and foreign television companies could mutually pool their advantages and avoid disadvantages.

Foreign media companies have many advantages compared to Chinese broadcasters, such as greater financial security, good human resources for management and marketing, advanced technology, strong innovation ability, as well as extensive international experience. However, they are not familiar with the Chinese market, media policy and the needs of Chinese audiences. In contrast, Chinese broadcasters do not have much know-how when it comes to producing high-quality programs; however, they know Chinese audiences and media policies much better.

### 6.3.3 Demand on Foreign Television Programs

Since the start of social modernization in 1978 the demand for foreign television programs in China has increased greatly, especially for feature films, television series, documentaries, sports programs, and music programs. Figure 7 shows that program imports steadily increased from 1993 to 2001. After the WTO accession the program imports became more important. The foreign program suppliers have gradually found their way to accessing the Chinese market. For example, the popular soap opera *Desperate Housewives* was screened on CCTV in 2005, which caused a great deal of excitement among foreign suppliers, because it was very rare for U.S. TV programs to find their way onto Chinese TV screens under the regime at that time which was strict about licensing foreign programs. (Bensinger, 2005, p. 7)

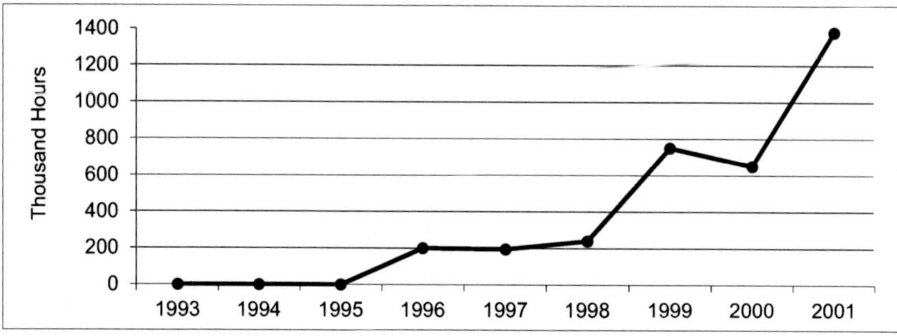


*Figure 7: Growth of Television Program Import in China (1993 – 2001)*

Source: CVSC Sofres Media, 2005.

There are several reasons for the huge increase in imports: (1) Television digitalization led to a great deal of unused television-channel capacity which must be filled by foreign programs; (2) increasing internet usage demands more foreign programs as video content on the websites (e.g., online TV series); (3) in the new television system the competition in the market forced Chinese broadcasters to provide a greater variety of programming; (4) cost pressure forced Chinese broadcasters to buy cheap programs on the market instead of investing in expensive self-productions.

*(1) Digitalization*

Market opening forced a technological advance. With the upgrade of the cable network to broadband, Chinese television has been progressively digitalized. The goal is to extend market capacity and increase the potential of economic development. (Guo, 2004, p. 14) Digital television could generate revenues not only from advertising, but also from pay TV and value-added services.

Digitization expanded channel capacity. More and more digital television channels were established. In 2003, 38 cable digital TV channels and 49 experimental digital cable systems were founded. By 2005, 30 million households received digital TV programs transmitted via satellite, and other 30 million households received digital signals via cable. (CCTV, 2002) By 2015 the analog-to-digital-television transition will have been completed.

A direct problem of digitalization is a lack of program content, especially in regard to high-quality programs. In 2001 alone the total television broadcasting time amounted to more than 9.5 million hours; however, only 2 million hours of programming could be produced by Chinese broadcasters themselves. (Fischer,

2003, p. 13) Consequently, program production capacity could not keep pace with the rapid increase in TV channels. The problem of a shortage of program content, which already existed in the analog era, was further reinforced by digitalization. A solution was for the capacity gap to be partially filled by imported foreign productions. Some Chinese broadcasters also expected to benefit from the know-how and financial means that foreign joint ventures would bring to the industry. (Guo, 2004, p. 22)

*(2) Internet usage*

The internet also developed very quickly in China. According to market research statistics, in 2005 there were 94 million internet users in China. This was an 8% increase compared to the previous year. The number of websites reached 669,000. (Zou, 2003, p. 2) Figure 8 demonstrates that the number of internet users increased faster than the number of television viewers from 1999 to 2005. On average every internet user spends 113 minutes per day on the internet (Zou, 2003, p. 2).

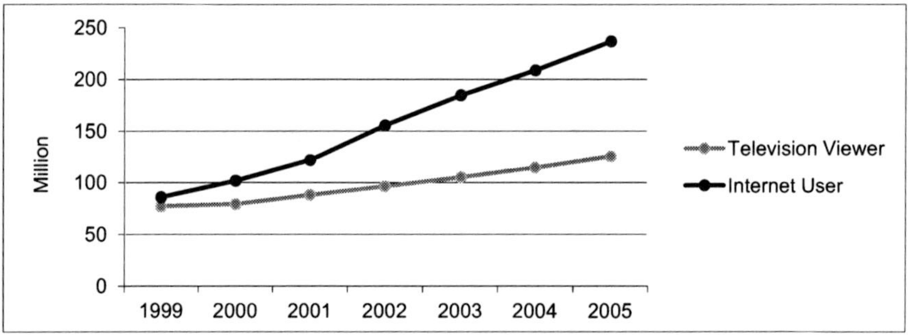


*Figure 8: Number of Internet Users and Television Viewers in China*

Source: adapted from Zou, 2003, p. 2.

*(3) Differential programming strategy*

As a result of the economic reforms a new social strata gradually arose in China with different tastes, needs, levels of education and income. Due to the pressures of competition, Chinese television broadcasters are required to broadcast a broad range of programs in order to satisfy the demands of various viewer groups. A differentiated audience market can serve the preference-oriented advertising market in an optimal manner, because advertisers are no longer only interested in big numbers, but rather in reaching the right audience. Thus, Chinese broadcasters need a large number of foreign television programs, which have a different style to Chinese programs in order to enrich the viewing choice for audiences and to attract certain audience groups.

*(4) Program purchasing in the market*

To promote competition in the Chinese television market and achieve greater economic efficiency, the separation of production and transmission was made of Chinese television. This refers to organizational separation on three levels: program production, compilation, and transmission. In other words, it refers to the outsourcing of program production and the transfer production resources to the market, because direct program purchasing from the market is much cheaper than internal production. Sometimes cheap programs purchased from the market can even appeal to a higher audience than expensive internal-produced programs.

*News programs*: After the reform the majority of news programs associated with propaganda are produced done in-house. "News production is strictly limited to those units that have a news producing license, i.e. *Xinhua News Agency*, *CCTV* and all other TV stations. News production is not allowed to be outsourced." (Fischer, 2003, p. 13)

*Non-news programs*: The non-news programs, such as mass-appeal films, television series, and sports programs, which depend less on propaganda, are produced by Chinese or foreign commercial suppliers. The government rarely intervenes in these areas and the voluntary choices between program suppliers and buyers take place in a decentralized way. The reform created an entertainment industry that is almost independent politically and has high degree of flexibility.

Only 28.8% of non-news programs were produced internally. 61.4% were purchased directly from other sources and 6.1% produced by external production companies under contract. The television broadcasters at lower administrative levels produced only about 10 to 20 hours per week, which covers less than 20% of their total broadcasting time. The rest of the broadcasting time is filled by programs from outside production companies (inclusive foreign programming) and programs from other broadcasters. (Ollig, 2002, p. 32)

# 6.4 Chinese audience

## 6.4.1 Demographic Difference

At the beginning of the 1980s after China had emerged from its destructive *Cultural Revolution* and had just restarted its modernization program through reforms and opening up (Xu, 2002, p. 6), many new phenomena regarding demographic differences surprised people. With economic development gender roles gradually changed based on changing traditional cultural values. A generation gap between people born before 1980s and after 1980s emerged due to the quite different living environment and value systems they experienced. Education opportunities were unequal for people in rural and urban areas, as well as those in economically poor compared to developed areas. The difference in income is large between the rich and poor in China.

*(1) Gender*

In the past 30 years gender roles have changed causing a transition from a male-dominated society to one in which females have started to ascend the social ladder. This is primarily reflected in the social functions and positions of men and women in the family, workplace, society, and education. Today, part of traditional culture still maintains some influence on gender roles in China, but economic development has also changed them significantly.

For centuries, Chinese social relationships and behavior between men and women was heavily influenced by Confucian ethical principles. Men typically had more power and status than women, given their differential role assignment within the family and society. Until only a few decades ago, only men could pass on the family name and inheritance, perform ancestral worship ceremonies, pursue educational and occupational goals, and become heads of nations. (Lau, 1996, p. 45) Men were the breadwinners and family heads who presided over the major decision making in the family. (Bond, 2010, p. 535)

The only place delegated to women was in the family where they played the instrumental and supportive roles of managing home and supplying male heirs. (Lau, 1996, p. 45) For women, their social behavior was governed by Confucian principles of three obediences and four virtues (*San Cong Si De*). These principles emphasized the obedience and subservience of women to their fathers when young, to their husbands when married, and to their sons when widowed. Women's social roles were narrowly defined within the family as producing offspring, serving and caring for family members, and keeping the households tidy and in order. (Bond, 2010, p. 535)

However, regarding the rapid social changes, contemporary China is not a completely male-dominated society without any opportunities for women. In most jobs it is more difficult for a woman to succeed than it is for a man. China is now trying to encourage its industries to give more opportunities to women. Although there are now some powerful female executives in the upper echelons of business in China, they are the exception. (Saxon, 2006, p. 65) The one-child rule has also helped greatly. If a family's only child is a daughter, the parents try to give her every advantage that they would have given a son, including higher education, which makes it easier for women to compete for higher-level jobs (Saxon, 2006, p. 65) and to attain higher status in society.

*(2) Age*

Due to China's *Reform and Opening-Up Policy* in place since 1978, the generations born before the 1980s and after the 1980s differ greatly, which has been referred to as *daigou* (the generation gap). More and more people recognized the increasing disparity between elders and youngsters in terms of attitude, behavior, and value judgments. (Xu, 2002, p. 6) The values of the youth in the 1950s and 1960s were generally consistent with mainstream social values, whereas in the 1980s Chinese youth seriously challenged orthodox norms as well as political and parental authority. (Xu, 2002, p. 6) Two factors are at work in creating a substantial generation gap in China today:

*China's one-child policy*: Chinese traditionally viewed the family as the center of their universe and the basic building block of their society. (Gundling, 2007, pp. 44-45) So children were customarily viewed as the greatest blessing of all, especially a family's only child. Consequently, there has long been a tendency to dote upon children. Chinese mothers were traditionally extremely protective of their children, more than in the Anglo-Saxon tradition that seeks to develop independence in children. (Gundling, 2007, p. 44)

*Rise in material wealth*: Before economic reforms most Chinese were extremely poor by Western standards. As a result, children could expect affection and nurturing in a family, but not many material belongings. (Gundling, 2007, p. 44) Since 1980 material wealth in China has risen quickly. Parents and grandparents are more inclined than ever to dote upon children. With their much greater disposable income they express their affection for children by buying them a lot of material belongings, e.g., toys, clothing, and so on. (Gundling, 2007, p. 44)

This so-called post-1980s generation is the first after economic reform in China and is considered spoiled, with the members of this generation dubbed "little emperors". They reflect nothing of past generations' willingness to work hard, save, and sacrifice. They have not "eaten bitterness" (*chi ku*) the way their parents did. They expect everything they want with no effort of their own required to get it. (Gundling, 2007, p. 44) All of which translates into their strong sense of optimism.

*(3) Education*

The education opportunities are sometimes unequal for men and women in China. In traditional Chinese culture the parents' decisions regarding a child's education are dependent upon the rate of return from child education. For instance, if the return to education for boys is better than for girls, due to gender discrimination in the labor market, then parents may feel justified in investing more in boys. (Chu, 2010, p. 138)

There are also rural-urban differences in education. Under financial pressure from the central government, since the 1980s the policy of educational decentralization has gradually been implemented, which renders local governments responsible for most of the expenses of their schools, including primary, junior high, and high schools. A consequence of this is that the local government in poor areas cannot afford to run many schools. Even though since 1980 some new schools have been established in rural areas, the school enrollment rate in there is obviously lower than in urban areas. College education is beyond the reach of most rural children anyway. (Chu, 2010, p. 138)

The income level of the family also has great impact on the education achievement of individuals. This effect became evident in the 1990s, which is consistent with a gradual worsening of local government finances. (Chu, 2010, p. 138) Therefore, children's education is closely related to how much money a family could afford to spend on it. Children from rich families could go to college, and even abroad to study. In contrast, children from poor families have financial difficulty just paying primary school fees.

*(4) Income*

The transition from a planned economy to a market economy brings changes to the distribution of income across sectors, regions and households. There was a growing consensus among observers that the benefits of reform were not equally distributed throughout China. (Wen, 2001, pp. 58-59) During the reform period rural and urban areas displayed a large and increasing inequality in income. Recently, the degree of income inequality has indeed been relatively high, when compared to China in the past or to the general level internationally. (OECD, 2010, p. 111)

## 6.4.2 Regional Difference

*(1) Different degree of economic development*

A country of 9.5 million square miles cannot grow at the same rate in all regions. The effect is an increase in the disparity of the rates of regional economic growth between eastern, central, and western China. There are several reasons behind the fact that the economic growth of central and western China is trailing far behind east China:

*Preferential government policies in eastern regions*: These are designed especially to attract foreign companies to invest in eastern regions. (Wijgers, 2008, p. 3) The opening up of China to the outside world has proceeded gradually from one region to another. Throughout the 1980s and the first half of the 1990s most of China's open areas were found along the coast, which became foreign investors' first choices thanks to preferential policies and a favorable investment environment. As most investors are engaged in export-oriented industries, the coastal areas are convenient outlets for their products. (OECD, 2002, p. 61) Thus, the rapid international trade growth since 1978 exacerbated interregional inequality. The richest 15 provinces, which accounted for 43% of China's population, generated 75% of exports, while the poorest nine provinces accounted for only 8% of exports, yet contained 37% of the population. (Wen, 2001, p. 60)

*Geographical characteristics*: The natural conditions of central and west China are harsher than those of coastal east China. West China, in particular, is covered by highlands, deserts and snow-clad mountains that are not inhabitable and not arable. The farming conditions there are very harsh compared to east China. This is also the case with regard to the development of transportation, telecommunication services and other infrastructure. (OECD, 2002, p. 61)

*(2) Different degree of openness*

Based on regional differences of economic development, the degree of openness, education, income, development of the media in costal east China is much better than in central and west China. People in east China have more opportunities to come into contact with people from other cultures in daily life and have more intercultural experience. Their minds and attitudes toward other cultures are more open. In contrast, in remote areas in central and west China, people do not have any contact with the outside world. They do not have a telecommunication infrastructure. Many people have never seen a telephone or a television in their lives. Therefore, in general, people in central and west China are more conservative and follow traditional Chinese culture more.

*(3) Different regional cultures*

Within China there are large cultural differences between regions. (Björkstén & Hägglund, 2010, p. 119) China has one of the world's oldest and most complex cultures. In large geographical regions in China the customs and traditions vary greatly between towns, cities, provinces, and regions, such as communication style, lifestyle, food, clothing, rituals, habits, and so on. For instance, people in north China like noodles, people in south China like soap, people in east China like sweet dishes, and people in central China like spicy food.

## 6.5 Limitation of China Studies

Many studies were conducted about the media in China by both, Chinese and Western scholars. They have some obvious limitations:

(1) Television in China was researched primarily as a political issue. In the center of interest is often the organization of media control and political manipulation. Scholars also paid a great deal of attention to change in role of Chinese television during phases of political liberalization in China. Chinese literature about television is mostly descriptive and does not contain any exact conclusions. In general, empirical research in this area is very rare. Information regarding data is often incomplete and unreliable. Even official dates from the state bureau are not up to Western standards. They are often arbitrary and inconsistent with each other (Staiger, 2003, p. 721).

(2) Many Western scholars have tried to research television in China using a Western scientific approach. Because of financial, personnel and communicative problems the statistical data collection in China is difficult. Furthermore, the size of samples is always a problem. Due to the heterogeneity of a huge population, a representative study is almost impossible. In some studies the samples were always concentrated in an area in a city and thus regional differences were rarely considered. As discussed, regional differences are very large in China and scholars should perhaps pay attention to people in different regions.

(3) Some influential factors which may influence research results are seldom used as control factors, such as demographics and television viewing. For example, differences might exist between different groups of people, for instance those who are highly educated and those who have a low level of education, or those who watch a lot of television and those who watch little. It is also of interest to scholars to investigate differences between audience groups to certain degrees of representativeness.

# 7 Design of Case Study in China

## 7.1 Research Object

The research object of this study is *cultural difference in foreign TV programs from the perspective of Chinese viewers*. This research object consists of five important aspects. In following each aspect will be interpreted in detail:

(1) *Cultural difference* in this study means the perceived difference between the background culture in foreign programming and viewers' own host culture. Real-life situations could be shown on television which reflect cultural difference, for example, weddings, friends visiting, and birthday parties. In these real-life situations viewers can see greetings, kissing, clothing, rituals, living environments, and so on. Many of these are culturally determined and can vary from culture to culture. Thus, they can be misleading and misunderstood.

(2) *Foreign country* includes all countries in the world except China, such as the United States, England, Germany, Japan, and Korea. Television programs from Hong Kong, Taiwan and Macau are not classed as foreign TV programs.

(3) *Foreign television programs* not only include foreign programs broadcast by Chinese TV stations, but also those downloaded from the internet or watched online. They conclude various television genres, e.g., movies, TV series, sports programs, documentaries, entertainment shows, and news programs. They have three different production forms: (a) the programs, which are produced by foreign media companies alone; (b) the programs, which are co-produced by both foreign and Chinese media companies; (c) the programs, which are produced by Chinese media companies, but in foreign program formats.

(4) *China* indicates five sample cities with different cultural characteristics in China – Beijing as the political center, Shanghai as the economic center, Guangzhou as a port city for international trade, and Nanjing as well as Xi'An as two cultural cities with long histories. The selection of these five sample cities is based on two reasons:

Firstly, they have different degrees of openness and control. For example, the culture in Xi'An is more conservative than in Shanghai and Guangzhou. Viewers living in the capital city Beijing are more heavily exposed to propaganda, because the control of television content in Beijing is the strictest in the country. In contrast, viewers in Guangzhou, which is a city near Hong Kong used for international trade, are reached by many Western media sources.

Secondly, due to the large population and various regional cultures in China it is not possible to consider China as a whole. There may be some generalizations that people can make about Chinese culture and society, but they must recognize that there are numerous regions (the East, West, North, and South China) and within each region there are huge varieties in terms of lifestyles, belief structures, attitudes, values, and so on. Although the results in the five sample cities could not offer generalizations about the whole of China, they could consider regional differences to a certain degree.

(5) Cultural differences *from the perspective of Chinese viewers* are used here to distinguish cultural differences interpreted by media corporations and control institutions. Chinese television viewers are viewers who reside in China.

## 7.2 Research Aims and Methods

In this study a new methodological approach is developed to examine cultural difference. In the center of interest is the perception and acceptance of cultural differences in foreign television programs from the perspective of Chinese television viewers. Unlike many other descriptive studies, this study is based on empirical methods and aims to gather reliable and systematical data.

The starting point of this study is the discussions about cultural discount and universal media culture. Hoskins proposed that a particular television program rooted in one culture will have a diminished appeal elsewhere, as viewers will find it difficult to identify with an unfamiliar or unexperienced culture in foreign programs. Some other scholars proposed that television programs, which are produced for the global market, create a universal media culture, in which audiences have mutual cultural knowledge and a sense of belonging to a global community. A result of this is cultural imperialism, which is frequently used to sum up the domination of American media products.

To examine both theories two questions must be answered:

*(1) What concrete cultural-difference types could cause a diminished appeal of foreign programs?*

*(2) To what extent do viewers find it difficult to identify with some unfamiliar or unexperienced culture in foreign programs? In other words, to what extent could viewers perceive and accept various cultural differences in foreign television programs?*

Generally, this study aims to qualitatively explore various cultural-difference dimensions and then quantitatively measure the reception of each dimension. In five sample cities this study followed a three-stage approach to data collection: (See Figure 9)

*Stage 1: Television schedule analysis (Part Two: Pre-Study)*

Television schedule analysis as a pre-study aims to gather the most important genres of foreign television programs in China. This will supply the background knowledge for this study. This part is concerned with two questions: (1) What kinds of foreign television programs from which countries are broadcast in China? This question aims to build an overview of the supplied foreign television genres and their countries of origin. (2) What kinds of cultural adaptation are necessary when they are broadcast in China? This question aims to describe some kinds of cultural modification in foreign programs for reducing cultural barriers for Chinese viewers.

*Stage 2: Personal interviews (Part Three: Interview)*

Personal interviews as the first part of main research aim to qualitatively explore the most important cultural-difference dimensions in foreign programs based on the opinions of Chinese viewers. Unlike the cultural-difference dimensions of Hofstede, Kluckhohn, Parsons, and Hall, the cultural dimensions in this study are especially identified for foreign television programs. They are interpreted by Chinese viewers themselves and are thus understandable to them. All of the identified dimensions are summarized in a catalogue in order to build an overview of various cultural differences. Some dimensions might be related to television genres and some might not. Then, the cultural background of each cultural dimension is explained, which offers information about the cultural reasons behind cultural difference.

*Stage 3: Survey (Part Four: Survey)*

A survey as the second part of the main research aims to quantitatively evaluate each cultural-difference dimension as well as to examine the impact of influential factors. In the survey, participants were asked to identify the degree of their perception and acceptance of each cultural difference between their own culture and the cultures presented in foreign television programs. *Perception degree* indicates how great a cultural difference they can perceive, while *acceptance degree* indicates to what extent they can accept cultural difference. Thus, the degree of perception and acceptance of cultural difference could be examined more precisely and a comparison between them is possible.

The part also aims to examine the relationship between the perception/ acceptance index and various influential factors, which may moderate the reception of cultural difference. They include demographics, resident region, use of foreign television, knowledge about foreign culture, and cultural proximity between the origin and destination countries. Due to a heterogeneous and huge population in China as well as limited financial resources, the representativeness of the samples could only be regarded as low degree in this study. This study pays some attention to regional difference in China, because regional differences of culture, economic development, openness, and lifestyle are large. By using cultural proximity as an influential factor, television programs not only from the U.S., but also from other countries are considered.

Both the personal interviews and survey are conducted on an individual level, because usage of foreign television programs occurs through individuals. This study pays close attention to the opinions of individuals who interpret and evaluate cultural difference within their own cultural frame. Based on the results of the interviews and the survey, Chinese participants are divided into various viewer types.

This study uses a mixed form of qualitative and quantitative methods, because each of the empirical designs used alone for examining cultural difference does not serve the results very well – each for a different reason. Qualitative interviews can explore cultural-difference dimensions and explain cultural background at a deep level; they cannot examine the validity of various opinions about cultural difference, because the selection of participants is usually random. The quantitative survey approach offers a substitute; it can examine the validity of each opinion to a certain degree, but through a survey alone people cannot know the cultural background of viewers' opinions about cultural difference.

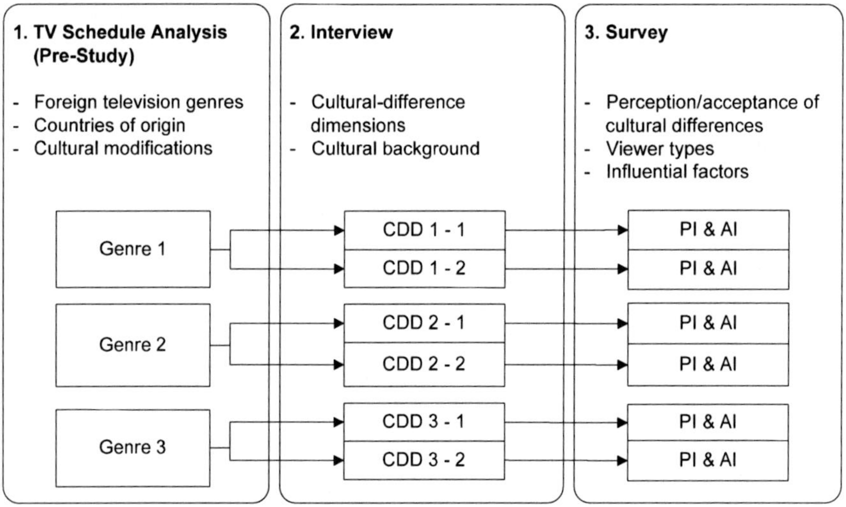


*Figure 9: Overview of Research Methods and Aims*
Note: CDD = Cultural-Difference Dimension; PI = Perception Index; AI = Acceptance Index
Source: Own illustration.

# Part D: Empirical Results

# Part D1: Pre-Study - Television Schedule Analysis

## 8 Types of Foreign TV Programs in China

### 8.1 Research Purpose and Research Questions

The first stage of researching viewers' opinions about cultural difference is to find out what kinds of foreign television programs are actually broadcast in China. In other words, what kinds of foreign programs can Chinese viewers really watch? Based on the answer to this question an overview can be constructed of the foreign programs supplied in China.

The most suitable methodological approach for answering this question is the television schedule analysis. Television schedule summarize various kinds of information about all broadcast television programs in a list that includes the date, start/end time, television channel, title of particular programs, and a short description of the programs.

Three concrete research questions in the pre-study are:

(1) *What kinds of foreign TV programs are broadcast in China?* This question aims to summarize the genres of foreign television programs in China, which will be used further in the personal interviews and the survey.

(2) *What kinds of cultural modifications must be made in order to fit the habits and tastes of Chinese viewers?* This aims to examine the extent of various necessary cultural modifications in foreign programs, such as content modification, language modification, and person modification.

(3) *How similar are the cultures of the country of origin and Chinese culture?* This aims to examine whether the foreign programs, which are broadcast in China, come primarily from culturally proximate countries which have a similar culture to China.

### 8.2 Method

A code book was created for coding various criteria for each program and consists of two parts: (See Appendix 1: Code Book)

The first part includes the program number, program name (e.g., *Animals World*), city (e.g., Beijing), TV channel (e.g., *CCTV-1*), start/end time (dd-mm-yyyy hh:mm), and duration in hours.

In the second part all of the variables were further divided into three groups:

*(1) Content catalogue – genre, theme*

This variable group codes programs according to genre and theme. The code book contains seven television genres: 1 = movie, 2 = television series, 3 = sports program, 4 = documentary, 5 = music program, 6 = entertainment show, and 7 = news program. Each genre has various themes. For example, the genre "movie" has 12 different themes: action, adventure, crime, romance, science fiction, and so on. "Television series" have 5 themes, "sports program" has 24 themes, "documentary" has 9 themes and so on. If a program cannot be classified into a genre or theme, it was coded as "0".

*(2) Cultural modification – content, language, person*

This variable group codes the kind of cultural modification and to what extent it is conducted for each program. There are three measure groups in this part: content, language, and person modification.

*Content modification* is measured in four levels: "1 = foreign content without content modification" means that the content of the foreign program is completely original without any changes to it. The second level "2 = foreign content with some content modification" means the program has been re-edited so that it caters to the habits and tastes of Chinese viewers. The third level "3 = foreign content with some added local content" means that the foreign video material has had Chinese material added in order to create a new program. The fourth level "4 = local content with a foreign format" means the program format has been imported from a foreign country, but that the content was produced in China. For example, the format of the dating show *Take Me Out* has also a Chinese vision.

*Language modification* is measured in four levels: 1 = original foreign language without subtitles, 2 = original foreign language with subtitles, 3 = dubbing, 4 = replacement of original spoken dialogue. The first three levels are easily understood. The fourth level "replacement of original spoken dialogue" means that due to problems understanding it, the original spoken dialogue in foreign programs must be replaced by new spoken dialogue created by the Chinese television presenters or editors, because the ways of explaining certain things can be quite different in diverse cultures. For example, the commentary for foreign football matches in China is always done by Chinese presenters.

*Person modification* is measured in three levels: 1 = no person replacement, 2 = some person replacement, 3 = complete person replacement. Person replacement means that the people in foreign programs, such as actors, presenters, or participants, must be replaced by Chinese. For example, in the co-produced German entertainment show *Wetten, dass..?* the Germany presenter and game participants must be replaced by a Chinese presenter and Chinese participants in order to attract more viewers in China.

*(3) Cultural proximity – country of origin, cultural similarity*

First, each foreign program was coded with a code for country of origin. Then each program was given a code for cultural similarity. This indicates how close Chinese culture is to each foreign culture. The countries considered as having a "1 = very similar" culture are Japan and Korean. The countries having a "2 = similar" culture are Thailand, Vietnam, and other Asian countries. Countries with a "3 = different" culture are (a) the United States, Canada, and other countries in North America, (b) the United Kingdom, Germany, France, and other countries in Europe, (c) Australia, and New Zealand from Oceania. The countries with "4 = very different" cultures are Brazil, Argentina and other South American countries, as well as countries in Africa.

Here it should be noted that cultural similarities are extremely varied and they are difficult to generalize. In this study the classification can be only made by using very general dimensions: geography, history, cultural origin, and language. Primarily, in this study the most important criterion to adjudge cultural similarity is Hall's concept of high- and low-context culture, which describes cultural proximity imprecisely. Therefore, cultural similarities here should be taken as a general guide for further research, rather than as iron-clad rules.

The samples of television schedules were selected according to television broadcaster, television channel, and date. A total of 37 television channels from eight television broadcasters in five sample cities were selected as samples (see Table 5). Sample days consist of an artificial week in March 2009 (see Table 6).

*Table 5: Sample Television Broadcasters and Channels in Five Cities*

| **TV Broadcaster** | **TV Channel** | **N** |
|---|---|---|
| *(1) Beijing* | | |
| China Central Television | CCTV- 1, 2, 3, 4, 5, 6, 7, 8,10, 11, 13, 14, 15 | 13 |
| Beijing Television | BTV- 2, 3, 4, 6, 10 | 5 |
| *(2) Shanghai* | | |
| Shanghai Media Group | SMG- 5, 6, 7, 8, 12, 13 | 6 |
| *(3) Guangzhou* | | |
| Guangdong Television | GDTV- 3, 6 | 2 |

| | | |
|---|---|---|
| Guangzhou Television | GZTV- 3, 6 | 2 |
| *(4) Nanjing* | | |
| Jiangsu Broadcasting Corporation | JSBC- 3, 4, 5, 6 | 4 |
| Nanjing Broadcasting Group | NJBG- 2, 3, 5, 7 | 4 |
| *(5) Xi'An* | | |
| Shanxi Television | SXTV- 7 | 1 |
| Total | | 37 |

Source: Own illustration.

*Table 6: Sample Days in March 2009*

| Monday | Tuesday | Wednesday | Thursday | Friday | Saturday | Sunday |
|---|---|---|---|---|---|---|
| 02 | **03** | 04 | 05 | 06 | **07** | 08 |
| 09 | 10 | **11** | 12 | 13 | 14 | **15** |
| 16 | 17 | 18 | **19** | 20 | 21 | 22 |
| **23** | 24 | 25 | 26 | **27** | 28 | 29 |

Source: Own illustration.

## 8.3 Results

The results of the television schedule analysis indicate that foreign programs (615.3 hours) occupied a significant part (13.2%) of the total broadcasting time (4662 hours) of the 37 sample channels across seven days. In the following, the percentage of total air time given to each television genre will be given.

The results also indicate that every sample channel has a certain time window during which it broadcast foreign TV programs every week. For example, *CCTV-3 Arts Channel* broadcast foreign documentaries about nature and animals in the time slots 06:30 - 07:00 and 16:00 - 16:00 from Monday to Friday. *CCTV-8 Television Series Channel* broadcast five episodes of foreign television series at night every day from 22:00 - 02:30. *CCTV-14 Children Channel* broadcast foreign cartoons from 12:00 - 13:00 and 21:00 - 22:00 every day.

### 8.3.1 Genres of Foreign Television Programs

The results indicate that the broadcast durations of various foreign television genres are quite different. The genres are divided into three groups according to the air time they received: much-supplied genres, little-supplied genres, and not-supplied genres (see Figure 10).

- The *much-supplied* genres are sports programs with 196.7 hours (32.0%), television series with 189.3 hours (30.8%), and documentaries with 141.8 hours (23.0%), because they are relatively culture-neutral.

- The *little-supplied* genres are movies with 37.7 hours (6.1%), music programs with 34.2 hours (5.6%), and entertainment shows with 15.6 hours (2.5%).

- The other genres, such as foreign news and talk shows, are not supplied in China.

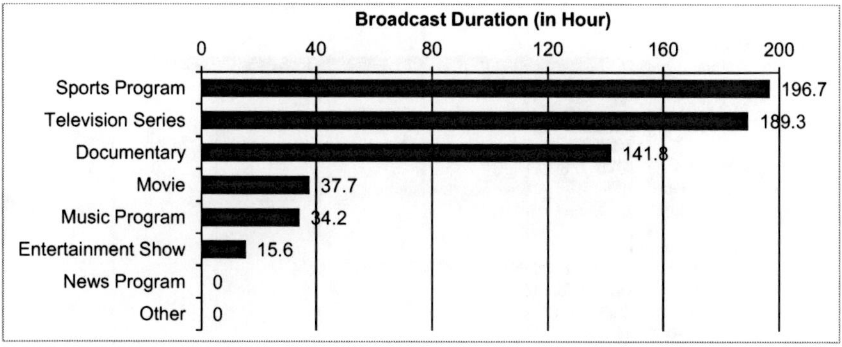


*Figure 10: Genres of Foreign Television Programs*

Source: Own illustration.

The themes of each supplied foreign television genre are varied (see Figure 11). The themes with less cultural content, such as "cartoon" as a theme of television series (104.8 hours), "basketball" as a theme of sports programs (94.9 hours), and "nature and animal" as a theme of documentary (56.4 hours), are broadcast more than other themes for each genre. In the following, the themes of each genre will be interpreted in detail.

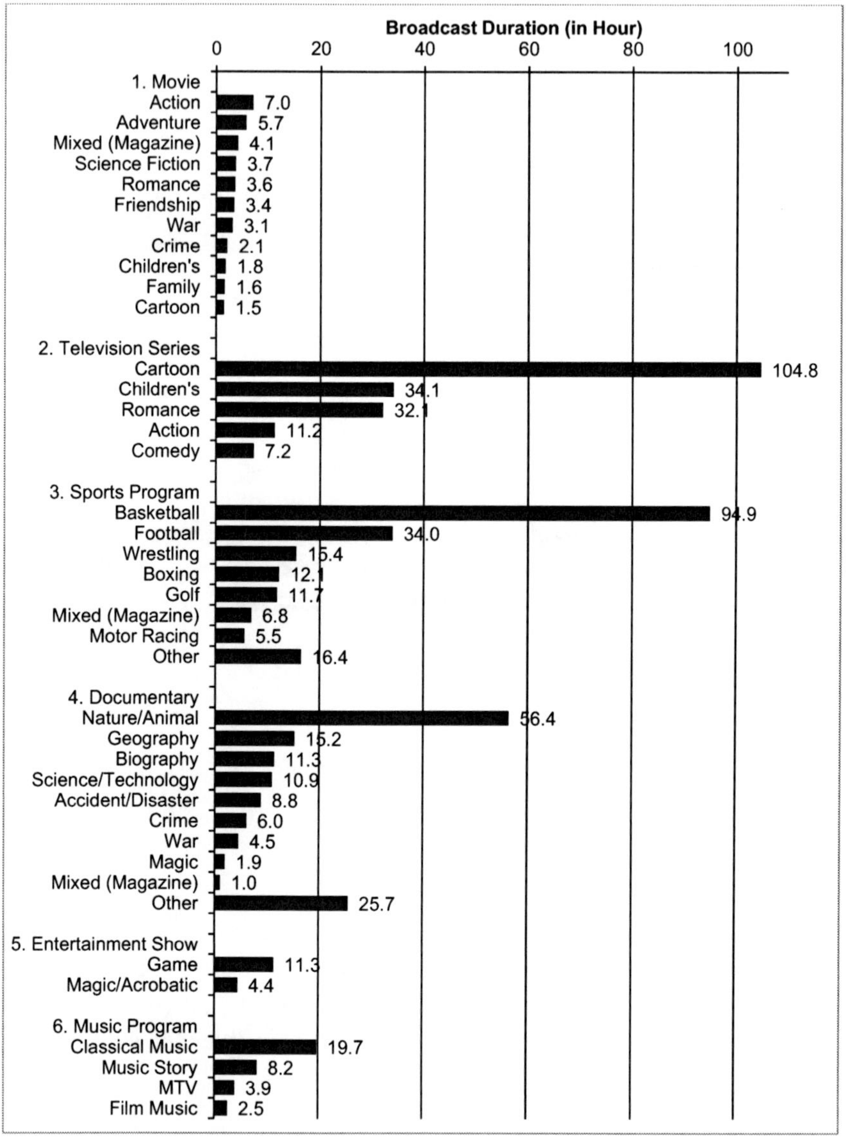


*Figure 11: Themes of Foreign Television Programs*

Source: Own illustration.

- **Movie**

Foreign movies in the sample schedules are primarily Hollywood blockbusters (e.g., *Spider-Man*) and classics (e.g., *Wuthering Heights*, *Scent of a Woman*). The themes of foreign movies are varied, such as adventure, action, science fiction, family, children's, crime, war, and romance. Action (7 hours), adventure (5.7 hours), and science-fiction movies (3.7 hours) are broadcast the most because they are relatively less related to culture than other themes, such as family movies (1.6 hours). (See Figure 11)

It is important to point out that there are some television magazine shows about foreign movies that are produced by Chinese media companies. These magazine shows use clips from movies to produce new television programs that showcase foreign movies. For example, the *China Central Television* produces two television programs *The Travel of the World's Films* (ShiJie DianYing ZhiLü) and *Expo of the World's TV & Film* (ShiJie YingShi BoLan), which use the best clips to present newly released or the most popular foreign movies from around the world. Each movie is discussed for about 10 to 20 minutes. The presenter of both programs is Chinese. She explains the plot of foreign movies in Chinese in such a way according to the habits of Chinese viewers. This Chinese way of explaining foreign movies could attract more Chinese viewers who have problems understanding foreign movies or who do not want to spend the time needed to watch an entire foreign movie.

- **Television Series**

Television series are one of the most broadcast foreign television genres with 189.3 hours, which is just a little less than sports programs with 196.7 hours in first place, but obviously more than documentaries with 141.8 hours in third place. (See Figure 10)

Figure 11 demonstrates the air time given to various themes of television series. The themes of foreign television series are relatively homogenous. The main type is cartoons with 104.8 hours, which sees this theme as the dominant one among all five themes of television series. One reason is that they present a fantasy world that whisks audiences away from everyday life (Rohn, 2010, p. 99). For example, the children's channel of *China Central Television* (CCTV-14) has a program called *Cartoon World* (DongMan ShiJie), which broadcast foreign cartoon series in the 12:00 - 13:00 and 21:00 - 22:00 time slots every day. Cartoon series broadcast in China include some Western ones (e.g., *SpongeBob*, *Thomas and Friends*, *Disney Cartoon World*, *Hans Christian Andersen: Fairy Tale and Stories*) and some Japanese ones (e.g., *Detective Conan*, *Doraemon*, *Sakura Momoko*).

Television series for children (34.1 hours) account for a relatively small portion of the total broadcast time. They include, for example, *Fimbles* from the British Broadcasting Corporation (BBC) and *Dragon Club* from Disney. Both programs are television magazines which broadcast various foreign television series for children.

The other three themes are romance (32.1 hours), action (11.2 hours), and comedy (7.2 hours). They include the Philippine romance series *Commitment* (ChengNuo), the Spanish action series *Zorro*, and the Korean comedy series *Funny Family* (GaoXiao YiJiaRen).

- **Sports Program**

Sports programs are the most broadcast genre with 196.7 hours. (See Figure 10) They include sports magazines, sports club programs and so on, produced by foreign media corporations. They do not include live sports events such as the football *World Cup* or the *European Champions League* because such sports transmissions are not typical in cultural context produced television programs. The themes of sports programs broadcast in China are quite diverse covering a broad spectrum of sports types. (See Figure 11)

Programs about basketball are broadcast the most (94.9 hours) and clearly dominated among all sports types. Basketball programs include *NBA Production* (NBA ZhiZao), *Basketball Park* (LanQiu GongYuan), *Basketball Camp* (LanQiu DaBenYing). Programs about football stay in second place (34.0 hours). They include *World Football Magazine* (ShiJie ZuQiu ZaZhi) and *Champions League Magazine*. Wrestling (15.4 hours), boxing (12.1 hours), golf (11.7 hours), and motor racing (5.5 hours) are less attractive for Chinese viewers. However, they are still important areas of foreign sports programming. There are still some magazines that feature a mixture of sports (1 hour), e.g., *Gillette World Sport* and *Global Sport* (TiYu ShiJie).

These sports programs are more or less tailored to Chinese viewing habits. For example, in some programs the most exciting moments of sports events are edited together for viewers who do not wish to watch the entire event. The presenters of these foreign programs are always Chinese and they often add information in their moderation and commentary that will be of interest for Chinese viewers.

- **Documentary**

Foreign documentaries in the television schedule include *Discovery*, *National Geographic*, and *Saga* (Chuan Qi) which contain various themes. The most popular theme is nature and animals (56.4 hours) which dominates among the documentary themes. Geography (15.2 hours), biography (11.3 hours), science/technology (10.9 hours), and accident/disaster (8.8 hours) are also an important part of the programming schedule. The other themes such as crime, war, and magic are rarely broadcast. (See Figure 11)

Similar to movie and sports magazines, foreign documentaries are also often broadcast in magazine format using Chinese presenters. For example, *China Central Television* has two prominent programs *Animals World* (DongWu ShiJie), which has been on the air since 1981, and *Man and Nature* (Ren Yu ZiRan), since 1994. They both show foreign documentaries concerning nature and animals. In the wake of their success, numerous others have started to use foreign documentaries about nature as program content, for example, *Green Space* (LüSe KongJian), *Wild Animals* (KuangYe DongWu) from *China Central Television*, as well as *Insight* (ShiYe) and *Military Affairs* (TieXue JunShi) from the *Shanghai Media Group*. Foreign documentaries about science and technology are broadcast in *Encyclopedia Quest* (BaiKe TanMi), *Technology Expo* (KeJi BoLan), and *Science World* (ZouJin KeXue). There are still other documentary programs about biography and other diverse themes, such as *People* (RenWu) and *Around The World* (ShiJi GeDi) from CCTV.

- **Entertainment Show**

"Entertainment programs" and "entertainment shows" are two different concepts. The entertainment programs, for example, movies, sports programs, and documentaries all have an entertainment function for viewers. However, entertainment shows here mean game shows, talent shows, variety shows, and so on.

Compared to the other five genres the foreign entertainment shows are broadcast the least, receiving only 15.6 hours of air time. Entertainment shows also have only a few themes. Game shows, with 11.2 hours, include *Wetten, dass .. ?*, *Survivor*, and *Robot Wars*. (See Figure 11) Usually foreign entertainment shows have to be localized for the Chinese market, because the original shows contain a lot of foreign cultural elements that do not fit the habits of Chinese viewers. Normally only the format is sold to China, and the Chinese television stations produce the programs using local personnel and content but in the foreign format.

For example, *Wetten, dass .. ?* is a famous entertainment show in Germany. In this program people challenge themselves and compete against each other in games. The participants and presenters are German. This program also has a Chinese vision which is co-produced by German and Chinese media corporations. In the Chinese vision the game participants are not only German but also Chinese, and the presenter is Chinese. The game rules are also partially changed.

Besides *Wetten, dass .. ?* the foreign dating show *Take Me Out*, the talent show *American Idol*, and reality show *Survivor* also have Chinese versions.

Another theme is magic and acrobatics (4.4 hour), an example of which is the *Monte Carlo Acrobatics Festival* broadcast without any obvious modifications. Some Chinese stations use clips from foreign entertainment shows to produce their own programs, such as CCTV's *International Art* (GuoJi YiYuan) which presents diverse foreign arts performances, comedy sketches, and so on.

- **News**

Although there are no foreign news programs in schedule, they represent a very important genre for researching cultural difference. In China foreign news programs are not allowed to be broadcast by cable TV channels. People in China can only watch foreign news programs via satellite in hotels, a few research institutions, and some resident areas for foreigners, because satellite receivers are generally prohibited for the Chinese population. They can also watch foreign news programs through the internet. The foreign news programs, which the Chinese viewers are familiar with, are *CNN News* (Cable News Network) from the United States, *BBC News* (British Broadcasting Corporation) from Britain, *Les Guignols de L'info* from France, and *Asahi Shimbun* from Japan.

- **Music Program**

The foreign music programs are broadcast relative rarely (34.2 hours) and occupy fifth place. (See Figure 10) Like other genres, foreign music programs must be more or less localized in China. They use clips from foreign music programs or music videos, but the presenters of these music programs are always Chinese.

For example, CCTV has three music programs, which often use foreign music material as program content. One is *Television and Film Phonograph* (YingShi LiuShengJi). This program uses famous and popular music clips from foreign movies and television series, such as the theme songs from the movies *Charlie and the Chocolate Factory*, *Sister Act*, and *Seven Brides for Seven Brothers*. The second is *Music Story* (YinYue GuShi). Its content often comprises stories about world-famous musicians, such as *Violin Geek – Nigel Kennedy*, *Ricky Martin*, *Movie Music Master – Ennio Morricone*, and *Coldplay*, which are produced by foreign media companies. Another is *Classic* (JingDian), which shows classic foreign songs, such as *Country Music*, *Natural Voice – Africa Hymn*, concerts and so on.

The Chinese version of *Music Television* (MTV) has been broadcast in China since 1995 and is produced by *Viacom Media Networks*. Its content is primarily foreign music videos and the presenters are Chinese. This music television channel now has various programs broadcast in China, such as *MTV Village* (MTV TianLaiCun), *MTV Top Music* (MTV PaiHangBang), and *MTV Music Ceremony* (MTV YinYue ShengDian).

- **Regional difference**

The results of the television schedule analysis indicate that the provision of each television genre in each sample city is quite different. Television stations in Beijing are the dominant supplier of various genres. The breakdown of the total output for TV stations in Beijing by genre is as follows: 25.9% television series, 24.0% sports programs, 18.4% documentaries, 15.3% movies, 12.8% music programs, and the remaining 3.7% are entertainment shows. The provision of TV stations in Shanghai, Guangzhou, and Nanjing is concentrated around television series, sports programs, and documentaries, which account for over 95% of broadcasting time of all foreign television programs. In Xi'An, 85% of foreign programs are sports programs and the remaining 15% is entertainment shows. (See Table 7)

*Table 7: Provision of Foreign Television Programs in Five Cities*

| | Beijing | | Shanghai | | Guangzhou | | Nanjing | | Xi'An | |
|---|---|---|---|---|---|---|---|---|---|---|
| | Hour | % | Hour | % | Hour | % | Hour | % | Hour | % |
| Sports program | 56.6 | 24.0 | 37.7 | 24.1 | 40.1 | 40.1 | 36.3 | 39.4 | 26.1 | 85.0 |
| TV series | 61.3 | 25.9 | 47.0 | 30.1 | 44.2 | 44.2 | 36.9 | 40.1 | - | - |
| Documentary | 43.5 | 18.4 | 70.0 | 44.8 | 10.8 | 10.8 | 17.5 | 19.0 | - | - |
| Movie | 36.1 | 15.3 | 1.6 | 1.0 | - | - | - | - | - | - |
| Music program | 30.3 | 12.8 | - | - | 3.9 | 3.9 | - | - | - | - |
| Entertainment | 8.7 | 3.7 | - | - | .9 | .9 | 1.4 | 1.5 | 4.6 | 15.0 |
| News | - | - | - | - | - | - | - | - | - | - |
| Total | 236.5 | 100 | 156.3 | 100 | 99.9 | 100 | 92.1 | 100 | 30.6 | 100 |

Source: Own illustration.

### 8.3.2 Cultural Adaptation

Television programs transmitted across borders cannot escape from some forms of transformation and cultural adaptation due to differences in language and culture. Even the world's most marketable U.S. feature films require at least an adaptation into different languages through subtitling and translation. Although many foreign television makers have endeavored to produce programs with uniformity for all audiences across the world and extended content appeal in order to increase their overseas acceptability, some cultural modifications must be made in order for the programs to better fit the viewing habits of local audiences. For example, South Korea orients its TV series for export and makes them more acceptable to wider Asian audiences. In China, foreign television programs also need some cultural modifications for them to be more understandable and acceptable for Chinese viewers.

For a variety of cultural, economic, religious, or political reasons, these processes of cultural modification involve interventions by social institutions and television broadcasters, and relate to factors such as production, marketing, scheduling and censorship. (Cohen, 1992, p. 23) First, some social institutions are actively involved. From the start and throughout all stages of the modification process, there is an organized and institutionalized mediation in place, which is operated by license, censorship, and so on. In China, the regulations for foreign television programs by the government are very strong and significant. The regulators closely examine program content and make any necessary modifications, e.g., remove or modify prohibited content. What might or might not be omitted depends on whether the content is negative or runs counter to the state's guidelines. Second, television stations themselves must also make some cultural modifications to foreign programs in order to tailor them for domestic viewers.

Each type of cultural modification can be measured at different levels. At the *minimum level* a program crosses borders without undergoing any changes. At the *maximum level* a program goes through many changes prior to reaching its potential audience. The *medium level* lies in between the aforementioned extreme levels.

Cultural modification levels vary between fiction and non-fiction programs. Fictional programs, for the most part, are broadcast in their original form with only minimal modification in the host country. Active social intervention does take place as in the case of linguistic translation, scheduling, and processes of selection; but it does not occur often. Thus, it can be said that fictional programs lie closer to the minimum modification. In contrast, non-fiction news programs tend to reside closer to maximum level and require censorship as well as modification to a high degree. (Cohen, 1992, p. 28)

However, the main burden of cultural adaptation is influenced by television viewers, because viewers interpret messages in programs according to their own knowledge, social situations, life histories, and worldviews. Besides all of other social and political reasons adapting foreign programs, the most basic criterion of cultural modification is that they are understandable and acceptable for domestic audiences.

Cultural adaption, which could be examined through a TV schedule analysis, includes three aspects: content modification, language modification, and person modification. Each modification type has three to four levels. Figure 12 demonstrates how many hours of foreign television programs are culturally modified at various levels. In the following, each modification type is introduced in detail.

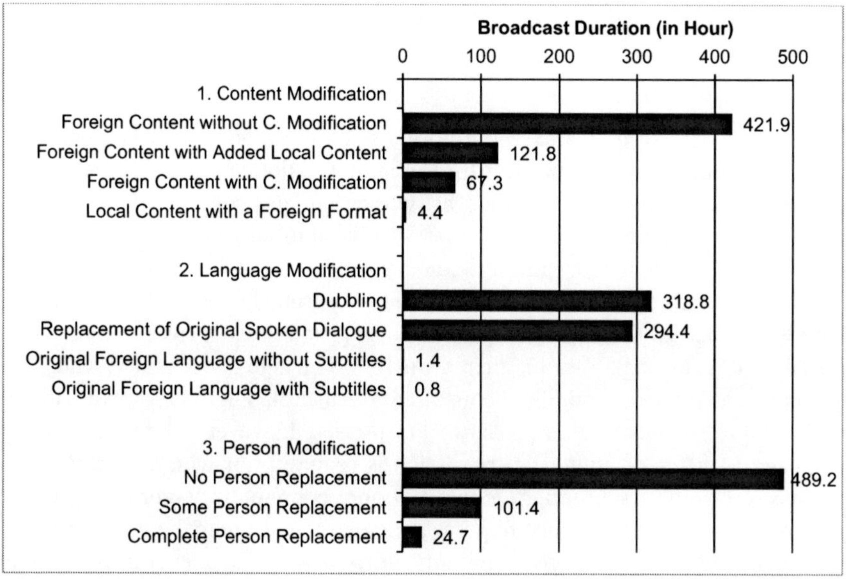


*Figure 12: Cultural Modification of Foreign Television Programs*

Source: Own illustration.

- **Content Modification**

Content modification means that the original foreign programs are more or less newly edited in ways that make them more relevant and meaningful to an audience in the destination country. With content modification, Chinese TV producers use foreign programs as "raw materials" to create new programs. A foreign program goes through the entire process of selection and editing until the Chinese editor feels it is suitable for broadcast in China. First, they split a foreign program into several parts and select or omit some of them. Then, they create a new program with a local presenter and comprising parts of the foreign television programs. Sometimes, they also add local content to the new programs.

Content modification has four levels: (1) foreign content without modification (minimal); (2) foreign content with some modification (medium); (3) foreign content with some added local content (medium); and (4) local content with a foreign format (maximal).

Figure 12 demonstrates that most foreign television programs are not modified in terms of content and remain at the first level. Their total air time (421.9 hours) is obviously longer than that for modification at the second (67.3 hours), third (121.8 hours), and fourth levels (4.4 hours).

- **Language Modification**

Language modification is the second form of cultural modification. Language is one of the biggest cultural barriers for foreign television programs. Thus, almost no foreign programs could be broadcast without at least some level of language modification. What is of concern here is what kind of language modification occurs if a program is produced in one country and broadcast in another.

A distinction should be drawn between two groups of importing countries, based on whether the exporting and importing countries speak the same language. If the language is same, then virtually no change is necessary. Thus, for example, an American program being aired in Australia might be shown without any changes. (Cohen, 1992, p. 25) In most cases, however, where the spoken language is different, linguistic translation is required, for which there are two options: dubbing or subtitling. *Dubbing* is more common in larger countries, as it is more expensive and hence requires a larger audience to make it economically viable. *Subtitling* is more commonly utilized in smaller countries, as it is much cheaper. However, it requires the audience to read a considerable amount of text. (Cohen, 1992, p. 25)

The language modification has four levels: (1) original foreign language without subtitle (minimal); (2) original foreign language with subtitle (minimal); (3) dubbing (medium); and (4) replacement of original spoken dialogue (maximal).

The results indicated that the language of almost all foreign television programs is modified significantly: 318.8 hours of programs were synchronized and 294.4 hours had new spoken dialogue added. Only 1.4 hours of foreign programming where broadcast in their original language without subtitles and 0.8 hours kept their original language with subtitles. (See Figure 12)

- **Personal Modification**

Personal modification means that the presenters, actors, or participants in foreign TV programs must be replaced by local people, for example, presenters of sports programs, actors in television series, or participants in entertainment shows.

Personal modification has three levels: (1) no person replacement (minimal); (2) some person replacement (medium); and (3) complete person replacement (maximal). The second level "some person replacement" means, for example, in sports magazine programs only the foreign presenter is replaced by a local presenter and other people (e.g., sports journalists) are not replaced. The third level "complete person replacement" is like in the case of the Chinese version of *American Idol*, where all foreigners including the presenter and participants must be replaced by locals.

The analysis results in Figure 12 show that people in 79.5% of foreign programs (489.2 hours) did not need to be replaced. People in 16.5% (101.4 hours) of programs were partly replaced, and only 4.0% (24.7 hours) of programs had complete personnel replacement.

- **Summary**

In Table 8 three aspects of cultural modification (content, language, person), each with three levels (minimal, medium, maximal), are compared. The results very clearly demonstrate that the content (68.6% at a minimum level) of and people (79.5% at a minimum level) in most foreign television programs did not need to be modified, whereas the language (51.8% at a medium level and 47.8% at a maximum level) of most programs had to be modified to a large degree.

*Table 8: Summary of Cultural Modification*

| | Minimal | Medium | Maximal |
|---|---|---|---|
| Content | 68.6% | 30.7% | .7% |
| Language | .4% | 51.8% | 47.8% |
| Person | 79.5% | 16.5% | 4.0% |

*Source: Own illustration.*

### 8.3.3 Cultural Proximity

Cultural proximity is an important concept in intercultural studies for examining cultural similarity between exporting and importing countries. In this part all exporting countries of foreign programs are ranked in terms of having small to large similarities with Chinese culture.

The results indicate that the TV programs from low culturally similar countries had total air time of 369.2 hours (80%), including the U.S. with 294.8 hours (63.9%) and European countries with 74.4 hours (16.1%). The programs from high culturally similar countries accounted for 91.9 hours (20%) of air time, including Japan 52.6 hours (11.4%), Korea 20.3 hours (4.4%), and other Asian countries 19 hours (4.1%). (See Figure 13)

This obviously shows that the programs from the United States, which has a low cultural similarity to China, dominate in China, whereas the programs from Asian countries, which have a high cultural similarity, only account for a small portion of the total air time for foreign programs. Thus, cultural proximity between exporting countries and China as an importing country is not directly related to how many programs are imported. More important is how well the content fits the preferences of Chinese viewers.

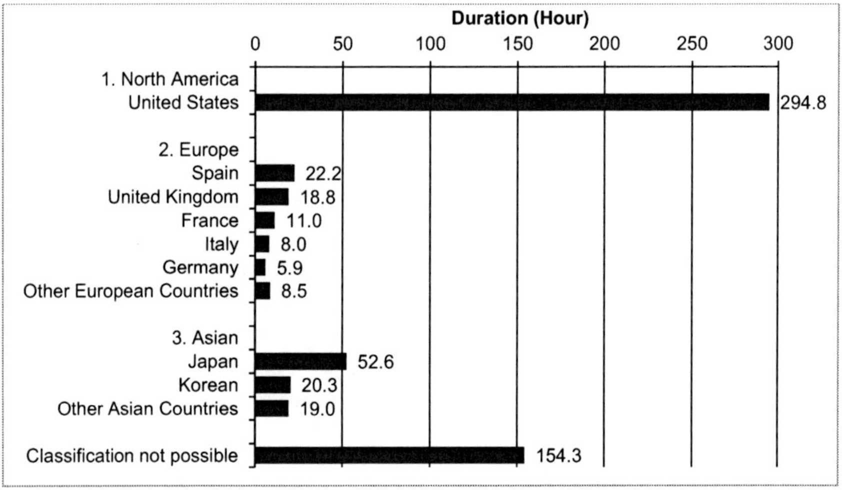


*Figure 13: Country of Origin*

Source: Own illustration.

Different countries of origin have different core areas of export. The United States dominated in all six television genres. Among European countries, Spain, the United Kingdom, France and Italy concentrate their exports on television series and sports programs. Japan, Korea and other Asian countries export primarily television series to China. The most exported Japanese TV series are cartoons. (See Table 9)

*Table 9: Cultural Proximity of Exporting Countries*

| | Movie | Series | Sport | Docu. | Music | Entert. | Total |
|---|---|---|---|---|---|---|---|
| U.S. | 25.2 | 61.0 | 126.8 | 69.4 | 6.4 | 6.0 | 63.9% |
| Spain | - | 11.2 | 11.0 | - | - | - | |
| U.K. | 1.9 | 9.5 | 6.0 | .5 | - | .9 | |
| France | 1.8 | 9.2 | - | - | - | - | |
| Italy | - | 4.0 | 4.0 | - | - | - | 16.1% |
| Germany | 1.5 | - | - | - | - | 4.4 | |
| Other Eu. | - | 4.2 | - | - | - | 4.4 | |
| Japan | - | 52.6 | - | - | - | - | |
| Korea | 1.6 | 18.7 | - | - | - | - | 20% |
| Other As. | - | 19.0 | - | - | - | - | |

Note: Series = TV series, Sport = Sports program, Docu. = Documentary, Music = Music program, Entert. = Entertainment show, Eu. = European countries, As. = Asian countries.
Source: Own illustration.

## 8.4 Summary of Part D1

This pre-study aims to answer three questions: (1) What kinds of foreign TV programs are broadcast in China? (2) What kinds of cultural modification must be made? (3) How similar are the cultures between the exporting country and China? Based on the answers to these questions, an overview can be made of the supplied foreign programs in China.

The pre-study selected and analyzed the television schedule from a total of 37 television channels from eight television broadcasters in five sample cities. The results indicate that foreign television programs accounted for a significant proportion of total broadcasting time. Every sample channel has a certain time slot for broadcasting foreign TV programs.

The first question summarized the *genres of foreign television programs* in China. Sports programs, television series, and documentaries are broadcast the most. Movies, music programs, and entertainment programs are broadcast the least. The other genres, such as news and talk shows, are not supplied in China.

The themes of each genre are not broadcast at an equal level. The themes with less cultural content, such as action and science-fiction movies, basketball and football programs, documentaries about nature and geography, as well as classic-music programs are obviously broadcast more than other themes, which have a great deal of cultural content.

The second question examined to what extent *cultural modification* must be carried out in order for a program to fit with the habits and tastes of Chinese viewers. The results show that the content of and personnel in most foreign television programs did not need to be modified, whereas the language of most programs had to be modified to a large degree.

The third question examined the *cultural proximity* between countries of origin and China. Programs from the United States dominated in all six genres, whereas the programs from culturally similar Asian countries accounted for a relatively small proportion of total foreign programming in China, being primarily television series. Some European countries concentrate their exporting on television series and sports programs.

# Part D2: Personal Interview – Cultural-Difference Dimensions

# 9 Cultural-Difference Dimensions in Foreign Television Programs

## 9.1 Research Purpose and Research Questions

Culture is that complex whole that includes knowledge, beliefs, art, morals, customs, and habits which are shared by members of a society (Yeshin, 1998, p. 307). Culture manifests itself in television programs. In many studies about cultural difference, language difference was often considered a major factor in determining cultural barriers in media products. However, language is far from the only factor which contributes to cultural barriers. Cultural difference in television may vary in quite different ways.

Television programs are multi-sensorial. Fundamental areas of television programs, such as colors, sounds, and symbols, may have totally different meanings for viewers resulting from their cultural interpretation. Most significantly, the cultural elements of television at a deep level, such as values and worldviews, result in markedly different attitudes towards foreign television programs.

Since cultural differences in TV programs depend on subjective perception and tastes, they are not easy to measure. To research cultural difference at a deep level is only possible when cultural difference types can be clearly defined and distinguished. Although numerous dimensions and notions of cultural difference in daily life were summarized by scholars in various studies, such as Hofstede's five cultural dimensions, they are not suitable for studies about television, because they are not interpreted by TV viewers. Therefore, to research cultural difference in television, firstly, various cultural-difference dimensions in television must be explored, which are understandable for viewers.

It is difficult to examine cultural difference between "Chinese television program" and "foreign television program". Instead, a more realistic way is to allow viewers to find a way to equate the "culture in foreign television programs" with "their own culture" that they are familiar with. In other words, how rational is the culture presented in foreign television programs compared to their own culture?

This study proposed that viewers have their own cultural values which stand in contrast with the cultures in foreign programs. Although each viewer can perceive many sets of cultural differences, some of them stand out because they somehow seem important or obvious.

Personal interviews with Chinese viewers were used as a method for exploring cultural-difference dimensions, because only the interpretations by Chinese viewers could be understood well by other Chinese viewers. The survey in the next research part could only be understood by Chinese participants when the terms and statements in survey were interpreted by Chinese themselves. Personal interviews aim to acquire the intuitive feelings of viewers – a more or less positive or negative attitude towards cultural difference in foreign television.

This part of the personal interview contains two research questions:

(1) *What kinds of cultural difference can be perceived by Chinese TV viewers*? This question aims to construct a catalogue in order to systematically summarize various cultural-difference dimensions based on the statements and interpretations of Chinese viewers, which will be further used in survey.

(2) *To what extent can they perceive and accept such cultural difference*? This question aims to categorize viewer types according to the extent they can perceive and accept cultural difference.

## 9.2 Method: Personal Interview with Chinese Television Viewers

### 9.2.1 Interview Conduction

The interviews comprised three phases: preparation, implementation and evaluation. Within the preparation phase the important definitions, research questions and interview guidelines were formulated. In the second phase, implementation, the interview guidelines were pretested, the sample cities were selected and all of the interviews were conducted. In the last phase, evaluation, all of the interviews were transcribed, and statements about cultural differences were selected, classified and summarized into a catalogue systematically, and all of the participants were divided into five types.

In China it is not possible to make appointments with interviewees several weeks or days in advance like in Western countries, because in China people tend not to stick to appointments. Thus, the author did not make any appointments with them. Instead, all of the interviewees were selected in person shortly before each interview was held. Posting letters is also not a good way to contact potential interviewees, because in China many people do not check their post box and may not even have one.

All of the interviews were conducted in book stores in the sample cities. In book stores, the author could find quite different interviewees with different genders, ages, educational backgrounds, occupations, and reading interests. The book stores were quieter than outside on the street, thus participants could better concentrate on interview and distractions could be better avoided. All of the book stores are located in the shopping center area in each city. They are:

- *Beijing Book Building* in Beijing (Chang An Road, Xi Cheng District);
- *Shanghai Book City* in Shanghai (Fu Zhou Road, Huang Pu District);
- *Guangzhou Book Shopping Center* in Guangzhou (Tian He Road, Tian He District);
- *Xin Hua Bookstore* in Nanjing (Han Zhong Road, Bai Xia District);
- *Jia Hui Han Tang Book City* in Xi'An (Chang An Zhong Road, Yan Ta District).

All of the interviews were conducted according to the interview guidelines, but the order of questions varied in different interviews depending on the actual interview situation. Some parts of the interview guidelines were modified to allow a better flow of the conversation. The guidelines ensure that the opinions of different participants could be compared with each other more easily. Before interview was conducted the interview guidelines were pre-tested with five Chinese participants.

Each interview lasted 10 to 30 minutes. The duration of the interviews depended on how well the interviewees could talk about the topic. In such a limited amount of time it was not possible for every participant to explain all kinds of cultural difference. The interviews were recorded with a voice recorder and later transcribed.

The interview guidelines were divided into three areas according to the core research questions:

*(1) Viewing foreign TV program*

They were firstly asked: "What foreign TV programs do you watch on television or over the internet, such as foreign movies, television series, cartoons, documentaries, sport shows, entertainment shows and news programs?" This was asked first because it is easy to answer and helps participants remember familiar foreign programs that they have seen. (See Appendix 3: Usage of Foreign Television Programs)

*(2) Perception and acceptance of cultural difference*

The next questions were: "Could you perceive any cultural differences in the foreign TV programs you have watched? Could you give some examples?" First, participants could say something freely about cultural difference. If they did not have any idea about cultural difference, the interviewer would then give them some interpretative framework to help them state something about cultural difference. Then they were asked: “To what extent could you accept such cultural differences?”

*(3) Demographics of participants*

At the end of the interviews the participants were asked to divulge some general personal information including their age, education level, and resident city.

### 9.2.2 Sample Description

All of the interviews were conducted on Fridays, Saturdays, and Sundays from October to November 2009 in China (see Table 10). A total of 36 Chinese TV viewers from the five sample cities (Beijing, Shanghai, Guangzhou, Nanjing and Xi'An) were personally interviewed. The purpose of selecting the different sample cities was to seek possible different opinions about cultural differences and to take into consideration possible regional difference.

*Table 10: Date of Interview Conduction*

| Sample City | Date | Day |
|---|---|---|
| Nanjing | 10 – 11 October 2009 | Saturday, Sunday |
| Xi'An | 23 – 25 October 2009 | Friday, Saturday, Sunday |
| Beijing | 30 October – 01 November 2009 | Friday, Saturday, Sunday |
| Shanghai | 07 – 08 November 2009 | Saturday, Sunday |
| Guangzhou | 28 – 29 November 2009 | Saturday, Sunday |

Source: Own illustration.

In the present study all of the interviewees were selected according to gender, age, education level, and residence city. The total interview sample size was 36. 58.3% percent of participants were male and 41.7% female. Their ages ranged from 15 to 50, with 69.4% between the ages of 20 and 29. Their education level ranged from primary school to Ph.D., with 52.8% having bachelor graduation. (See Table 11) Their occupations were lawyer, scientist, designer, teacher, worker, student, and so on. They had different interests in terms of books, such as foreign culture, Chinese literature, world travel, religion, media, sports, movie, and history. (See Appendix 2: List of Interviewees)

The different demographics, occupations, and reading interests of participants could help to ensure that different opinions were gathered. However, it should be pointed out that due to the small number of participants, the representativeness of the sample and comprehensiveness of all the cultural difference types cannot be guaranteed in the research results.

*Table 11: Sample Description*

| | | BJ | SH | GZ | NJ | XA | N | % |
|---|---|---|---|---|---|---|---|---|
| Gender | Male | 4 | 3 | 3 | 5 | 6 | 21 | 58.3 |
| | Female | 3 | 2 | 3 | 3 | 4 | 15 | 41.7 |
| Age | Under 20 | 0 | 0 | 0 | 1 | 2 | 3 | 8.33 |
| | 20-29 | 3 | 2 | 3 | 3 | 4 | 25 | 69.4 |
| | 30-39 | 3 | 2 | 2 | 2 | 4 | 13 | 36.1 |
| | 40-49 | 1 | 1 | 1 | 2 | 0 | 5 | 13.9 |
| Education | Primary school | 0 | 0 | 0 | 0 | 1 | 1 | 2.80 |
| | Middle school | 1 | 1 | 0 | 2 | 1 | 5 | 13.9 |
| | Bachelor | 4 | 2 | 2 | 6 | 5 | 19 | 52.8 |
| | Master | 2 | 1 | 4 | 0 | 2 | 9 | 25.0 |
| | Doctorate | 0 | 1 | 0 | 0 | 1 | 2 | 5.60 |
| Total | N | 7 | 5 | 6 | 8 | 10 | 36 | 100 |
| | % | 19.4 | 13.9 | 16.7 | 22.2 | 27.8 | 100 | |

Note: BJ = Beijing, SH = Shanghai, GZ = Guangzhou, NJ = Nanjing, XA = Xi'An
Source: Own illustration.

# 10 Results I: Cultural-Difference Dimensions

The first part of the interview results deal with what types of cultural difference could be perceived by Chinese participants and what their cultural backgrounds are. In the interviews all of the participants said that in foreign television programs there certainly had some cultural differences. They described their problems or difficulties when watching foreign television and also engaged in a cooperative interpretation process about the reasons they sometimes could not understand foreign programs and had barriers to accepting them.

The interviewees' statements underlined the fact that their common understanding of cultural difference consists of many dimensions. These statements form an overview of several sets of cultural differences. Some are related to certain television genres; in other words, each television genre has its own special cultural-difference type, while others are not related to television genres.

It is important to point out that participants overwhelmingly described their opinions in positive terms. However, they also held embedded negative judgments about certain cultural barriers they had experienced. People reacted differently to different aspects of cultural difference. This seems typical of the mixed nature of intercultural experiences. In the following, various cultural-difference dimensions from the perspective of Chinese viewers are interpreted in detail.

## 10.1 Openness vs. Conservatism as a Cultural Boundary

Almost every participant in the interviews has a very strong Chinese cultural identity, which gives them a solid foundation and set of standards for adjudging cultural phenomena. They do not question whether there are any cultural differences in foreign television programs. Rather, they have a very clear boundary line drawn in their minds between "our" and "their" culture, and they have two frameworks for interpreting and adjudging them both.

In the interviews, "open" and "liberal" were the two words used most often to describe cultures in foreign television programs, in contrast to the "conservative" culture in China. The interviewees classified the culture in foreign television as being one of openness and liberty, whereas Chinese culture leaned more towards conservatism. Almost every interviewee used this criterion as a boundary for distinguishing both cultures.

The openness and liberty of the culture in foreign programs is reflected in various aspects, such as the behavior and personality of the characters, and their way of thinking, education methods, moral standards, and living habits (see Table 12). In the following, each aspect of this boundary will be introduced in detail based on the statements given by the interviewees.

*Table 12: Openness vs. Conservatism as a Cultural Boundary*

| **Culture in foreign TV programs** | | **Chinese culture** |
|---|---|---|
| Open/liberal | | Conservative |
| ← | Censorship of TV content | → |
| ← | Personality of characters | → |
| ← | Values | → |
| ← | Moral standards | → |
| ← | Education methods | → |
| ← | …… | → |

Source: Own illustration.

- **"Their" culture is open, "our" culture is conservative**

> "I think their culture is more liberal and our culture is a little inflexible and conservative." (Interview XA_10)
>
> "I think the Eastern and Western cultures have major differences. The thinking of foreigners is relatively more open, whereas the thinking of Chinese is more conservative." (Interview NJ_6)

"Their culture" and "our culture", "Eastern and Western cultures" indicate that in the minds of the interviewees there is still a clear boundary between Chinese culture and other cultures. The most significant difference is the degree each are perceived as "open" and "conservative". They consider Western culture to be typically open and liberal, whereas Chinese culture is conservative.

In the opinion of several interviewees the difference between open and conservative cultures is reflected primarily in the *censorship* of television content. One important reason is that Chinese television has a long tradition of propaganda (Shirk, 2001, p. 91). The Chinese government believes that propaganda as an educational approach was more effective than the use of force or punishment in controlling people's thinking and thus they used it to promote what they believed to be correct. (Zhao, 2000, p. 258) This approach is based on a Confucian belief that people tend to imitate moral paragons (Bai, 2007, p. 103).

The propaganda function of Chinese television is controlled by the government's strict regulation. No program is free from censorship. (Hong, 1998, p. 46) Program content must correspond exactly to the policies and goals of the government. They are not allowed to threaten social stability or challenge the party's worldview (Zhang, 2009, p. 115). Censorship guarantees that inappropriate program content, which would have a negative effect on the government and society, has no chance of being broadcast.

Wang (an advertising agent from Shanghai) thought that "the attitudes towards some sensitive themes, such as religion" between Chinese culture and foreign cultures in television were very different. She considered censorship in China to be "strict", whereas in foreign country it was "loose". She said:

> "The control of television content in foreign countries concerning religion, sex, violence, morality and so on is relatively loose. In contrast, in China the censorship of these sensitive areas is very strict." (Interview SH_3)

She was also very aware that foreign programs imported into China had been modified according to political guidelines prior to broadcast by "some government departments".

> "Some foreign television programs imported into China first have to be changed according to the criteria governing censorship. Through censorship, some content will be removed and some foreign programs will even be banned in China." (Interview SH_3)

In contrast, some of the interviewees believed that in Western countries (e.g., the United States and European countries) the content of television programs is controlled little by the government. Television makers in these countries have the freedom to produce programs that viewers really want to watch and which could contribute to democracy and opinion diversity in society. Wu (a manager from Beijing) said: "Foreign countries do not have a propaganda department similar to China's, because our system of television regulation is different to other countries" (Interview BJ_7). Furthermore, Wang explained why television programs must be strictly controlled in China:

> "If Chinese television programs were as open as foreign ones, then the values and standards of morality of many people in China would be strongly influenced. So Chinese television programs are not allowed to be very open." (Interview SH_3)

This statement indicates the Chinese government's deep fear of chaos (luan) (Keane, 2004, p. 88). From ancient to contemporary China, *stability* has always been one of the most important issues in society. The Chinese are traditionally afraid of social instability. (Ambler, 2008, p. 43) In a country as heavily populated as China, stability whether political, social or economic has often been of great concern. The fear of chaos resonates, for many, from the events of 1989, but equally significantly from the *Cultural Revolution*, and the warring chaos of twentieth-century China. (Latham, 2001, p. 102) Due to the huge size of the country and its population, the highly negative consequences of such chaos should be avoided at all costs. As Wu said: "Chinese television programs cannot be shown without censorship" because "the population in China is huge" (Interview BJ_7).

If people in China were to receive too much negative information from television, which does not correspond to or even runs counter political guidelines, a direct consequence would be social instability. So the fear of chaos can be seen as crucial for maintaining the dominant position of the government in the television system. Thus, many non-media people from censorship departments give program makers suggestions for modifying program content in order to maintain government control of Chinese television (Hong, 1998, p. 48).

- **Openness and conservatism at surface and deep level**

The cultural difference of openness and conservatism is reflected in some aspects of television at a surface level, such as the personalities of the characters in television programs, which can be observed directly. In contrast, this cultural difference is also reflected in some aspects of television at a deep level, such as values, morality, and worldview, but which cannot be observed directly.

Generally, participants considered that people in foreign programs are "direct" and "open" (Interview SH_2). Their direct and open personality was primarily perceived from their speech. Zhang (an international trader from Guangzhou) said:

> "The Chinese presenters must pay attention to what they can say and what they cannot say. […] The foreign presenters on foreign television programs are more laid back. They can say whatever they want." (Interview GZ_3)

"Can say" and "cannot say" refer to the cultural difference of freedom of speech. In China, television presenters and commentators are prohibited from saying anything that goes against the government's guidelines. In contrast, in some Western countries presenters have more freedom of speech and can articulate their opinions in a more direct and open manner, because restrictions on speech are relatively more relaxed than in China.

The interviewees also thought that some cultural elements found at a deep level in foreign programs are more open than those in Chinese programs. The following statements are examples of such thinking: "the values in foreign programs are relatively more liberal and open" (Interview BJ_6); "their moral standards are more open than Chinese moral standards" (Interview SH_3); "in foreign programs we can see that education is more open than in China" (Interview GZ_2).

On the whole, they believe that the conservatism of Chinese culture both on the surface and at deeper levels is "deeply influenced by the Chinese traditional culture which has developed for several thousand years". (Interview SH_3)

## 10.2 Genre – Related Cultural Difference

In this chapter, the cultural-difference dimensions related to television genres are introduced separately. The six television genres are (1) movies & TV series, (2) sports programs, (3) documentaries, (4) entertainment shows, (5) news programs, and (6) music programs.

### 10.2.1 Movies and Television Series

The perceived cultural differences in foreign movies and television series (incl. cartoons) consist of three aspects: character, drama plot, and music.

- **Character**

The characters in movies and television series often have culture-related personality types, for example, *Rambo* as an American iconic hero. They are the products of culture and social development (Bachmair, 1996, p. 91). They are also intertwined with the experiences, imaginations and emotions of television viewers as they have a tendency to interpret movies or TV series in the form of a character or role (Bachmair, 1996, p. 121). Thus, character emerges from a culture, and the perception of it occurs within a culture.

The cultural differences of characters perceived by the interviewees firstly concerns characteristics of nonverbal communication. Although the language in television is undeniably important when communicating to viewers, equally important are the many nonverbal cues that accompany spoken messages. (Schmidt, 2007, p. 95) Nonverbal messages are transmitted outside of spoken language, but they also transmit a great deal of meaning and express emotions and attitudes.

Television viewers receive a variety of nonverbal messages from characters while watching television: personal appearance, kinesics or body movements (including facial expressions, eye contact), vocalic or paralanguage (includes volume, rate, pitch, and timbre) (Schmidt, 2007, p. 96).

These cues are culturally influenced. Understanding nonverbal communication helps viewers to better understand foreign characters. Characters in high-context cultures are more attentive to subliminal messages embedded in nonverbal behavior than characters in low-context cultures. So to avoid cross-cultural misunderstandings, nonverbal messages in programs, which are broadcast in other cultures, should mirror the cultural differences between the producer and viewers. (Schmidt, 2007, p. 100)

*(1) Appearance of a character*

Appearance of a character refers to physical appearance, clothing, costume, make-up as well as other symbols identifying the uniqueness of a character. Their appearance displays and communicates something to television viewers. Characteristics such as intelligence, sociability, trustworthiness, and success are initially determined by appearance. Status and prestige are also communicated by dress and appearance. (Schmidt, 2007, p. 97) Although these may be imprecise and unclear measures, they are important factors influencing a character's image.

Some interviewees thought "their appearance" has "surely" some differences as Chinese characters. However, in general, they did not consider the difference in appearances of characters to be great and they can "accept" it.

*Physical appearance*: The difference in physical appearance refers to the "race difference" of characters. Wang from Shanghai said that "the foreign actors or actresses have blond hair, blue eyes and they are taller than Chinese actors and actresses". (Interview SH_3).

*Costume*: Ding from Beijing said that "their ways of dressing sometimes look modern and strange" (Interview BJ_2). Zhang (a student from Shanghai) thought that "the attire worn by cowboys, such as jeans and cowboy hat" looks very exotic (Interview SH_4).

Zhang from Xi'An stated that he could accept the cultural differences of costumes in Western movies and television series, but he could not accept costumes in some TV series from other Asian countries. For example, he watched a Korean television series *Shen Ji Jian* (Supernatural Arrow) about the wars between Korea and China during the Ming Dynasty (A.D. 1368-1644), but it depicted Chinese solders wearing uniforms from the Qing Dynasty (A.D. 1644-

1911). His attitude towards this error was that “Korean television makers do not respect history and change it. I cannot accept this”. (Interview XA_10)

*Make-up*: Some participants thought that the make-up of Chinese actors or actresses looks too perfect and unnatural, because very thick foundations are used to cover blemishes on the skin. A schoolgirl, Feng, from Beijing said:

> “In Chinese movies people think that the more perfect the actors’ make-up is, the better. However, in foreign movies, the make-up of actors looks more natural. Sometimes we can still see some freckles on the faces of foreign actors or actresses.” (Interview BJ_6).

The difference in make-up is also related to Chinese culture. In China, white skin is considered more beautiful than dark skin (Gifford, 2008, p. 160). Thus, in summer many Chinese women use a parasol to prevent sunshine from tanning their skin. They also spend a lot of money on cosmetic products to make their skin look whiter. In contrast, in Western countries such as Germany, dark skin is seen as more beautiful, because brown skin looks healthy. Thus, some German women go to solariums to tan their skin and use cosmetic products to make their skin look darker.

*Criteria of beauty*: Aesthetic standards; in other words, what is considered beautiful varies between cultures (Rohn, 2010, p. 73).

On the one hand, some interviewees thought the criteria for judging beauty in Chinese culture and foreign cultures were the same. For example, Yu from Beijing said, “in modern dramas beautiful women and handsome men in Chinese television programs look similar to those in foreign ones” (Interview BJ_5). Wang from Shanghai stated that both, Chinese and foreign television program producers select actors and actresses who fit the roles on offer very well. Based on this criterion, they select actors and actresses who look beautiful, handsome and comfortable in front of the camera.

On the other hand, some others found that the criteria for judging beauty were different. “The difference is that the aesthetic taste of television viewers is different in each country. What a foreigner likes is different to what the Chinese like” (Interview SH_3). This interviewee used *Gong Li*, a famous Chinese actress in Hollywood, as an example:

> “Some foreigners think that *Gong Li* is a typically beautiful Chinese woman. Perhaps they like Chinese women of this type. However, we think other types of Chinese woman are more beautiful.” (Interview SH_3)

*(2) Personality of character*

Some types of nonverbal behavior particularly influence viewers' relational perceptions of characters' personality. Characters use body language to express and communicate nonverbally their feelings as well as attitudes with their facial expressions, gestures, posture, voice, and even silence. (Schmidt, 2007, p. 96) Smiling seems to communicate liking, composure, and formality. Eye contact reflects the cultural themes of hierarchy, social balance, and humility. A shrug of the shoulders may communicate helplessness or uncertainty. (Schmidt, 2007, p. 97) Television viewers could gather enough information from nonverbal behavior to establish an overall pattern that can reveal personality and identify the meanings behind facial expressions and gestures.

These nonverbal elements are cultural specific. For example, in the United States people use a lot more hand gestures as a sign of self-confidence (Wood, 2009, p. 130), but in China too much gesturing with the hands while speaking can be seen as offensive and as showing disrespect for other people (Sadri, 2011, p. 164). Sometimes Americans will rest their feet upon a desk and work in shirt sleeves in order to create an egalitarian working atmosphere, but Chinese would see this simply as lacking in formality and manners for (Huang, 1994, p. 240).

Some interviewees found that characters in foreign films and television series had different personalities compared to Chinese ones. According to them, the foreign characters' personalities were "obviously open", "casual", "free", and "direct". Thus, they had a "stronger individual personalities" than Chinese characters and looked "very self-confident" as well as "independent". In contrast, Chinese characters were relatively conservative and their behavior seemed more restricted and "implicit".

Fang (a student from Shanghai) said, "Their gestures are sometimes too exaggerated, especially when they are very happy" (Interview SH_5). Wang from Beijing said, "Their posture and facial expressions are sometimes exotic, such as those of the British gentleman" (Interview BJ_1). These statements indicate that they could perceive cultural differences in characters' personalities through nonverbal communication.

*(3) Acceptance of character difference*

The acceptance degree of foreign characters varied between participants:

- *Accept*: Some participants could accept foreign characters. Example statements are as follows: "Most characters are acceptable." (Interview XA_9) "There is no one who is not acceptable." (Interview XA_7)

- *Not accept*: However, others could not accept them if they looked too "exotic", "ugly", "absurd", "strange", or "open". "Sometimes characters are too ugly. If they are not too exotic then I can accept them." (Interview SH_5) "I cannot accept the very absurd characters" (Interview XA_2).

- *Like*: Some viewers like characters in foreign movies and television series because they think "It is interesting to see some cultural differences. If we always see the same faces, it is boring." (Interview GZ_1) Some others like characters in foreign cartoons because they look "very vivid".

- **Music**

Today music is a necessary and important element in movies and TV dramas because it is a stimulating factor for situations and offers emotional support to the action on screen. Music is able to create an atmosphere and it can help characterize individuals and their emotions. Good sound design can contribute to the positive perceptions of a movie or TV drama because pictures and voices alone are often not enough to do this. (Wolther, 2006, p. 252) Although listening to music does not require any cultural knowledge, the ways of perceiving and understanding musical information are dependent on different cultures because music can evoke associations with daily life as well as memories.

In the opinions of some Chinese interviewees the background and title music in foreign movies and TV dramas often have their own special cultural features compared to Chinese ones, in regard to quality, style, melody, and rhythm.

(1) *Music quality*: Yuan from Nanjing said that he liked foreign background music because "the quality of background music in foreign programs and Chinese programs is not at an equal level." (Interview NJ_1) She stated that foreign music was of a higher quality than Chinese music. Wang explained her attitude towards music quality: "foreign music regarding melody and mood is very good [...] because the music of many Hollywood movies is produced by some musical geniuses." (Interview SH_3) In contrast, she found music in Chinese movies and TV series to be "a little weak" in these areas. Wang also explained why she thought Chinese background music was not as good as foreign background music.

> "Usually foreign movies and television series use classical songs as background music. [...] Chinese pop music has developed relatively well. However, other Chinese music, such as Chinese classical music and folk music has not developed well, because not many Chinese like to hear it and the market for this type of music is very limited. So Chinese classical and folk music are also rarely used in Chinese movies and television series." (Interview SH_3)

(2) *Music style*: Generally, the interviewees found musical differences in foreign movies and TV series was "okay" and "not big". They could accept it if it sounded "normal" and not "exotic". However, some had difficulty accepting certain music that was too passionate or too rock, because they were used to hearing to milder music in Chinese movies and TV series. Ding from Beijing said: "I do not like rock music or music which is too exciting" (Interview BJ_2). Feng (a student from Beijing) explained that "if I hear some lively, noisy or vibrant music in foreign programs, I can feel it is foreign music." (Interview BJ_6)

In contrast to modern foreign music, others preferred classical music in foreign movies and TV series, such as symphonies or orchestral music. A schoolgirl liked classical music because it sounded "very elegant and can help me calm down" (Interview NJ_6). Zhang (a scholar from Xi'An) liked not only classical music from Western countries, such as the Beethoven symphonies, but also classical music from Japan, such as *Shen Si Zhe* (Interview XA_10).

- **Drama Plot**

> "I think foreign movies are more interesting than Chinese movies, because the plots of dramas and various other aspects (of foreign ones) are better." (Interview NJ_5)

The themes, proximity to daily life, and fantasy of drama plots were considered as three cultural-difference dimensions of movies and TV series.

(1) *Themes*: The statements of some interviewees indicated that they could perceive cultural differences in drama themes. They said that the dominant themes in Chinese movies and television series were more traditional than foreign ones, such as some historical and political themes about the *Cultural Revolution*, and the *Chinese Civil Wars*. These Chinese themes "do not have a modern flavor". Zhang (an international trader from Guangzhou) said:

> "The domestic television programs about war always tell stories about the internal troubles in China. People are shown fighting against and criticizing their own people, such as in stories set during the *Cultural Revolution* in China. I do not like to watch domestic television series of this type." (Interview GZ_3)

These statements indicated a cultural tradition of movies and television series in China. As mentioned in previous chapters, Chinese movies and television series not only have an entertainment function, but also one of propaganda to advocate patriotism and mainstream ideology in China. (Keane, 2004, p. 93) The protagonists of such programs were professionals who "prove" the ideology was correct. (Qian, 2008, p. 159)

In contrast, the themes of foreign movies and television series were more "modern" than Chinese ones. They covered science fiction, new technology, and so on. These themes were more "commercial" and correspond better to viewers' needs. Some young people in China like watching foreign movies and TV series of the themes science fiction, war, horror, thrillers, and they particularly like blockbuster movies. Some others could not accept violent and brutal ones, such as horror movies.

Wang from Beijing liked to watch American blockbuster movies because "they are produced with high computer technology" and they "look very real and very shocking" (Interview BJ_1). Zhang from Guangzhou also liked blockbuster movies, because "their special effects are very good, especially the visual impact they have in blockbusters" and "foreign television producers pay close attention to every detail" (Interview GZ_3). Yuan from Nanjing liked the foreign kung-fu movies because Chinese kung fu in foreign action movies uses modern visual effects and thus looks "better", "smarter", and "more impressive" (Interview NJ_1).

(2) *Proximity to daily life*: Although the themes of foreign movies and television series are much more modern than Chinese ones, some viewers preferred watching Chinese ones. One reason was that foreign drama plots were often fictional and far removed from their everyday life, whereas Chinese drama plots "reflect stories that are nearer to our lives"; in other words, they had more references and proximity to viewers' daily lives.

> "I feel the stories in many foreign television series are fantastical and unreal. However, Chinese domestic television series reflect our daily life and are proximate to real life in China." (Interview XA_8)
>
> "There is no benefit from watching foreign television programs. I think it is realistic to watch some Chinese programs, such as the Chinese programs dealing with what happens around us and in our daily lives. I know little about foreign television programs." (Interview BJ_3)

Proximity to daily life is particularly important for emotional or romantic movies as well as television series. Some Chinese participants would like to watch Chinese ones because "in foreign romance movies the way to express one's emotion is different to the way one would in China" (Interview SH_4). Thus, they "cannot think from the perspective of those depicted in foreign television programs" and they "can understand emotions in domestic television programs better" (Interview NJ_1). Jiang (a father of a four-year-old child) was concerned about the possible negative impact of foreign cartoons on his child, because "the content in foreign cartoons is not related to our life in China" (Interview NJ_7).

(3) *Fantasy*: Foreign movies and TV series, which were very fantastical, seemed very popular, "especially some cartoons, such as *Tom and Jerry*" (Interview BJ_1).

## 10.2.2 Sports Programs

> "I watch them with Chinese commentary. The foreign commentary is not easy to understand." (Interview XA_2)

(1) *Commentary*: In this statement commentary seems important when watching foreign sports programs. In the opinion of some interviewees, foreign commentators were obviously more critical than Chinese ones. "Critical" means that foreign commentators tended always to say something about scandals and the mistakes of sports players in order to attract more attention from viewers, while Chinese commentators tended to say more amusing things about sports players, sports events, and some stories about sports. They criticized sports players much less often than foreign commentators.

(2) *Commercialization*: Generally, Chinese viewers thought that foreign sports programs were more commercial than their Chinese equivalents. Zhang (a scholar from Xi'An) explained the reason for sports programs being commercial in the United States:

> "The whole of American society is very commercial, including the presidential, congressional, and state-governor elections. [...] So commerce penetrates various areas of American society" (Interview XA_10).

(3) *Sports culture*: Another cultural difference of sports programs concerns sports culture, which includes sports stars, slogans, corporate designs, and so on. In the opinion of interviewees foreign sports programs have their own special sports cultures and traditions, such as "professional sports clubs", e.g., the "NBA" and "professional boxing". Zhang said:

> "Although American sports programs are very commercial, they also have their own special sports culture, which is very good. [...] Although the Chinese sports programs are also commercialized to a certain degree, the feel of American and Chinese sports programs is different". (Interview XA_10)

(4) *Sports types*: The most preferred sports types in sports programs were also different, for example, boxing and wrestling were popular in the United States,

but they were rarely seen on Chinese sport shows. Furthermore, there were some special cultural sports on foreign sports programs, such as Spanish bullfighting; European equestrian and polo; Japanese baseball, sumo, and judo; as well as Indian cricket.

### 10.2.3 Documentaries

> "I think the cultural difference in documentaries is the broadest compared with other foreign television genres." (Interview BJ_7)

(1) *Objectivity vs. Subjectivity*: Some interviewees thought that foreign documentaries (e.g., about history, biography, and society) were "more objective" (Interview BJ_7) than Chinese ones, because they contained both positive and negative perspectives and had "fewer subjective opinions" (Interview NJ_4).

Zhang (a scholar from Xi'An) thought Chinese documentaries were sometimes "artificial" meaning that "Chinese documentary producers add or delete some facts intentionally". He cited a biographical documentary as an example: "Nobody is perfect. […] The Chinese producers just describe the positive side of a person, but neglect their negative side" (Interview XA_10). He stated that Chinese biographical documentaries always aimed to present a positive image of a person and thus came across as subjective.

However, Wu (a manager from Beijing) believed that this kind of objectivity of foreign documentaries was "relative", because "foreign documentary makers have surely a different agenda to ours. Their objectivity has a certain direction and must surely correspond with their own interests and guidelines" (Interview BJ_7). His statements indicated that although foreign documentary seemed objective, they must also follow certain rules and serve certain interests.

(2) *Explicit vs. implicit*: Explicit and implicit is a typical cultural difference between high- and low-context cultures, where the former places an emphasis on words whilst the latter is more reliant on contextual cues (Verderber, 2010, p. 51). This is also related to different preferences for directness and indirectness in media communication. (Cheng, 2003, p. 49)

Producers from low-context cultures (e.g., the U.S. and Britain) place emphasis on media communication being direct. Directness is a very desirable characteristic to display in media (Hunter, 2005, p. 209). Thus, their documentaries trend to be more logical and scientific. They provide enough background knowledge and provide evidence to substantiate the content as well as opinions given.

In contrast, Chinese media places emphasis on the indirectness, ritual, and "smooth" surface demonstration of media content. They prefer to tell viewers what viewers want to see instead of the absolute truth, because "the true lie" could sometimes make viewers feel better. If they want to inform viewers about problems or make suggestions to them, an indirect way would be chosen which seems more acceptable in China. Thus, Chinese documentaries tend to be more intuitive and appeal to viewers' emotions.

Some interviewees felt there the difference between explicit and implicit was "very large", because foreign and Chinese documentaries had two different frameworks or narrative styles for telling a story or interpreting something. Foreign documentaries tended to explain something explicitly, whereas Chinese ones always express something "implicitly". Zhang (an international trader from Guangzhou) said:

> "In some foreign documentaries, their subject, purpose, and structure are very clear. In contrast, Chinese documentaries do not describe things directly, but rather always implicitly. You need a long time to be able to understand [...] what they want to express." (Interview GZ_3)

(3) *Starting Point/Perspective*: In the opinions of the interviewees, the starting points and perspectives of foreign documentaries were different to Chinese ones. For example, Lin from Nanjing said, "their opinions and our opinions about the same things are different." (Interview NJ_4) Wu from Beijing explained: "when I watch some foreign documentaries about China, I can feel a big difference, such as their starting point to introducing and perspective on China." (Interview BJ_7) Zhang from Xi'An expressed his experience in this area. He watched two documentaries about China: one was produced by the *BBC* and was called *Wild China* and the other was produced by a Chinese television station. He thought both documentaries were completely different, although they had the same topic of China's nature and culture. He said:

> "If *Wide China* had been produced in China, it would only have stressed the positive side of China, such as China being a very good country and its social system being very good. Actually, *Wide China*, which was produced by the *BBC*, took quite a different starting point and had a different perspective to those Chinese producers would have adopted. [...] The *BBC* did not add any political leaning to this documentary." (Interview XA_10)

(4) *Theme*: Some viewers liked to watch foreign documentaries about "travel" and "animal protection". Wang from Shanghai said, "I like foreign documentaries about adventures or journeys around the world to investigate various cul-

tures. I have watched many shows like this." (Interview SH_3) In contrast, she stated that Chinese documentaries about similar themes were "very weak", because they do not have their own innovative ideas and just "imitate" foreign documentaries with themes about various cultures, customs, or nature. Wu (a manager from Beijing) said that many foreign documentaries covered "science" and "history" which had "education significance" for viewers, while some Chinese documentaries distributed "communism ideology". (Interview BJ_7)

### 10.2.4 Entertainment Shows

In entertainment shows, the presenter plays a very significant role of "connector", which means they must give background information to viewers during shows, which is needed to allow them be able to understand issues and the various opinions as well as get involved personally in topics. (Wachtel, 2003, p. 92) However, presenters of an internationally broadcast entertainment show cannot connect to both domestic and foreign audiences at the same time in equal measure. This might mean that a very popular presenter in a particular country is not accepted by foreign audiences in another.

Moderated entertainment shows, however, always have subjective components because they are based on one person's personality traits, such as speaking style (Oliver 2006, 330). The presenter's body language also speaks to the audience which is often clearer than verbal speech. Facial expressions and posture of the presenter support their opinions and also have an influence on their style of speech (Chimombo, 1998, p. 184).

*Speaking style of presenter*: In the opinion of some interviewees the biggest cultural difference between foreign and Chinese entertainment shows was the speaking style of presenters. Foreign presenters spoke in a more "open" and "casual" way. They had more freedom to articulate their own opinions. Sometimes they even made outrageous or controversial comments to grab the attention of viewers. In contrast, Chinese presenters were relatively conservative und appeared more serious. Wang (an advertising agent from Guangzhou) explained his experience in this area as the following:

> "Chinese presenters have to pay attention to what they can say and what they cannot say. They cannot say everything they want. Not only presenters, but also guests must be very careful about what they say. The foreign presenters in foreign television programs are more laid back. They can say what they want to. Sometimes it does not matter if they are impolite or even crude." (Interview GZ_3)

Generally, presenters and guests in foreign entertainment shows were considered to be more "direct" and "open" than Chinese ones. "Can say" and "cannot say" highlighted the cultural difference of speech freedom. In China, TV presenters and commentators are prohibited from saying anything that runs counter to the guidelines set down by the government and they have very strong awareness of what they should and should not say. This awareness is closely related to the education they receive before their carrier begins.

In China almost all presenters must receive strict training at special universities. They must learn the standard rules for television presenters. The most important part of their training is what they should and what they should not say. For example, they are not allowed to talk about sensitive themes such as political problems, sex, violence, or religion, and they cannot say any impolite or harsh words in television. Furthermore, they must learn the standard posture and gestures which give them the appearance of a typical Chinese presenter. Therefore, after their professional training many Chinese presenters lose their own personality when they are on television. They all look and behave similar to each other when presenting programs. Their speech always follows the rules they learned during their training.

It should be noted that in China what people can (not) say is not only strictly controlled in television programs, but also reflected in various areas of daily life (Damm, 2001, p. 69). For example, what teachers (not) can say to their students; what governor can (not) say to citizen; what people can (not) say to neighbors, relatives, or friends must be according to some cultural rules. In traditional education, Chinese are usually taught to be reserved and restrained. They are cautious about expressing their emotions and opinions. They usually do not express their opinions, demands, and criticism in a direct way; they rather use implicit words to let other people understand what they want to say (Mindess, 2006, p. 59). This is a typical phenomenon in high-context cultures – people understand each other in a cultural context.

## 10.2.5 News Programs

Chinese news programs have a tendency to avoid criticism, but stress positive topics, which aptly characterizes the socialist news approaches. (Straubhaar, 1992, p. 108) Chinese news programs almost never criticize or portray their governments, military, or economies as weak; rather they often praised their government and individuals for their social contributions. (Straubhaar, 1992, p. 107) China always portrays itself in news programs as having a strong tendency to cooperate with other countries and not competing, staying out of conflicts, or of being involved in dependent relationships. (Straubhaar, 1992, p. 100)

In contrast, the economy and the government are more often criticized by news programs in Western countries, reflecting the watchdog role adopted by the Western media. (Straubhaar, 1992, p. 108) For example, criticism of home and other governments was prominent in the United States, but praise for them was seldom found anywhere. (Hallin, 1994, p. 41) The party or government in power was often shown as wrong or incapable (Mauk, 2005, p. 297). Furthermore, the U.S. was most often reported as engaging in competition with other countries. (Straubhaar, 1992, p. 100)

*Diversity and sensitivity of news topics*: This is one of the biggest differences between Chinese and foreign news programs. In China, the themes concerning the government were touched upon as primary themes. Educational issues and food supply themes were covered more than Western countries. However, terrorism coverage was scant. (Straubhaar, 1992, p. 100-103)

In contrast, Western countries were revealed to be somewhat more "sensationalism" in their choice of news topics. The definition of what was sensational was often limited to accidents, disasters, and crime. Sensational political topics, such as demonstrations, protests, and criticism of the government, were more widely covered in American news. (Straubhaar, 1992, p. 98) Human rights were primarily a theme for Western industrialized countries, but not at all in China.

Generally, Chinese news programs are much more likely to cover solutions than problems, while the Western industrialized countries showed more problems than solutions, which reflected a critical, even negative approach.

## 10.2.6 Music Programs

In foreign music programs cultural differences are not as obvious as other TV genres. Zhou from Xi'An paid little attention to musical differences, because she cannot feel any difference in music (Interview XA_9). Wang from Shanghai liked pop music and classical music in foreign music programs, because both sound very nice (Interview SH_3).

Although "music is sometimes without country bounders", music programs have more or less some cultural differences (Interview XA_10). Zhang (a scholar from Xi'An) explained that "Eastern and Western music is a little different" in music programs, such as in terms of "music style".

## 10.3 Genre – Unrelated Cultural Difference

### 10.3.1 Language

Every language draws a border around people to whom it belongs. (Morley & Robins, 1997, p. 94) Each language has different critical sounds and learning a new language means learning to identify new sounds. Thus, people only hear the sounds of their own language which they have learned to listen to. Anyone would find it difficult to identify crucial differences in speech sounds in other languages. (Jandt, 1998, pp. 172–173)

If people grew up speaking English, the Chinese language would be difficult for them to perceive because some aspects of the Chinese language do not occur in English. So they do not know how to listen up for them and subsequently literally do not hear them. If people grew up speaking Chinese, there are also some aspects of English that are difficult for them to perceive, for example, English has some consonant sounds that do not exist in Chinese.

Language is very closely connected to society and culture. A language can be regarded as a social phenomenon; its systematic features or rules are dominated by the speakers of a language (Toepser-Ziegert, 1978, p. 8). Language is a necessary component of a culture, on the one hand, and a faithful reflection of this culture, on the other (Jedynak, 2011, p. 66). Thus the close relationship between language and culture may cause some cultural barriers for television viewers when they try to understand a foreign TV program in other languages.

Today the importance of language as a necessary element of television programming remains undisputed. Only through a combination of pictures and language can a television show be optimally presented. Television makers relay their information, opinions, and attitudes through the use of language (Bednarek, 2010, p. 121). Without language a television show is difficult to be understood completely (Seyfarth, 1995, p. 180). Language in television, however, is a complex and troublesome vehicle affecting both cultures and viewers involved in a meaningful exchange. Especially bothersome is the arbitrary and often abstract relationship between words, objects, and thoughts. (Schmidt, 2007, p. 101)

Intuitively, audiences are more receptive to content presented in their own tongue and more resistant to content in an unfamiliar language (Doyle, 2002, p. 93). Thus, linguistic obstacles have to be negotiated when TV programs are introduced to audiences who speak a different language. Although dubbing and subtitles are frequently as a means of translation, these presentational interventions disrupt viewing pleasure. (Fu & Sim, 2010, p. 122)

- **Subtitling and dubbing**

Paralanguage refers to everything beyond or in addition to language itself. Such variables as tone, rate, volume, breathiness, and nasality offer a great deal of information which can convey a speaker's personality traits, attitudes, status, and feelings. (Schmidt, 2007, p. 97) Thus, dubbing would change such paralanguage of the original language more or less (Varela, 1997, p. 319).

(1) *Original language with subtitling is more preferred than dubbing*: Some interviewees "only watch foreign television program in their original language with subtitles" (Interview BJ_6) because they "do not like the Chinese dubbing" (Interview BJ_2). Some of them explained why they liked and only watched the original versions. Zhang (a scholar from Xi'An) said:

> "If I watch the dubbed foreign programs I can just watch the emotional facial expression of foreign actors and actress, but I cannot feel any emotion from the Chinese dubbing. Therefore, I think the original language sounds better. For example, in 2003 I watched the movie *Titanic* in the original language. What Rose and Jack said and at the end of the movie what the old woman said was very moving." (Interview XA_10)

A total of five reasons were summarized as to why they like to watch foreign programs in the original language:

- Chinese dubbing sounded "strange", "uncomfortable" and "uncoordinated", because "Chinese voices cannot match foreign lip movements very well" (Interview BJ_2);
- Chinese dubbing "does not have the same feeling as the original voice" (Interview XA_2) and "sometimes dubbing cannot express the emotion of the original language very well" (Interview XA_10);
- Original voice heard "authentic", "comfortable", "pleasant", and "moving". Thus, "no Chinese dubbing could be better than the original language" (Interview SH_3);
- Dubbing is "sometimes quite confusing" (Interview BJ_7);
- They learned foreign language "through watching foreign television programs in their original language". They think this is a good way to learn a "pure" foreign language. (Interview XA_9)

Some of the viewers, who only watched foreign programs in their original language, did so without reading the Chinese subtitles, while others had to because their foreign language skills were "not very good". Zhang from Xi'An explained that if programs are in English, he does not need to read the subtitles. However, if the programs are in another language, such as Korean, he has to read them. Feng (a student from Beijing) stated that she could understand English just "a little". She found "it is more convenient to read the subtitles" (Interview BJ_6). Zhou from Xi'An believed that "although many Chinese can speak and understand English, they cannot speak English very well and they cannot completely understand English in foreign programs". Therefore, subtitles are "necessary and better than nothing" (Interview XA_9).

(2) *Dubbing is more preferred than the original language with subtitles*: In contrast, some participants found dubbed programs in Chinese to be better than those in the original language. They were "used" to hearing Chinese dubbing and "not used" to hearing the original foreign language. There were two reasons:

- They "cannot understand original language". (Interview NJ_5)
- They "do not need to watch subtitles". (Interview NJ_2)

(3) *Subtitles and dubbing are equal*: The other viewers did not care whether foreign programs were in their original language or were dubbed into Chinese. They accepted "both of them".

Firstly, whether viewers watched foreign programs in their original language with subtitles was primarily related to which foreign language they could understand. If they understood a foreign language very well and knew something about its related culture, they would prefer to watch these programs in the original language. Zhang (a student from Shanghai) explained that he needed language and intercultural knowledge in order to understand foreign programs:

> "At the beginning I was not used to hearing Chinese dubbing, but now I am. However, this is related to whether I know the particular foreign language well and its related culture. For example, I am used to hearing English, Japanese, Korean, and German." (Interview SH_4)

Secondly, the preference for original language or dubbing is related to television genres or actors. Wang from Beijing preferred watching American blockbuster movies with dubbing, whereas he preferred watching cartoons and some comedies in the original language (Interview BJ_1). Hu from Nanjing stated that his preference for original language or dubbing "is related to how well the actors or actresses act out their role". If they "act very well", he preferred the original language. If they did not, he preferred to hear Chinese dubbing instead. (Interview NJ_3)

- **Translation**

Translation as a linguistic phenomenon is located in the area between two linguistic systems and its task is to find the equivalent forms and content of one language in another (Toepser-Ziegert, 1978, p. 14). However, in the real world few words can find an exact equivalent in another language, because the words of different languages come from diverse socio-cultural contexts (Schopenhauer,

1992, p. 32). An individual's store of knowledge and social background in different cultural contexts prevent them from understanding a translation "absolutely".

Translation often makes it possible for a television program rooted in one culture to be broadcast in countries with different cultures. When a viewer reads a translation, every word, every punctuation mark, is not the author's, but the translator's. Translators act as mediators between people of different tongues (Tonkin, 2010, p. viii). Translation is defined in its simplest terms as an interpretation of verbal signs by verbal signs of another language. (Gonzalez, 1989, p. 484)

Equivalence between original text and translation is one of the key aspects of a good translation (Jandt, 2009, p. 137). It postulates a relation between the source-language and the target-language text. The content equivalence consists of stylistic, formal, functional, textual, communicative, and pragmatic equivalence. Thus, translators must decide what kinds of equivalence are important or not important from a hierarchy of equivalence values. (Gonzalez, 1989, p. 495)

Another criterion for a good translation is a translator's use of words that are common in the target language, but not in the source language. There are very few words in any two languages with identical emotional value and impact. Sometimes, the translator does not transfer individual words in their context, but rather thoughts and ideas, which can be conveyed very well in the other language. (Gonzalez, 1989, p. 488) A translation should reproduce the meaning and intent of the original text correctly and completely, affect the audience the same way the original does, and not sound like a translation; that is, it should sound like an original work. (Gonzalez, 1989, p. 485) This is regarded as proof that the translator knows his own language.

Also, by watching foreign television some problems regarding viewers' understanding of translation may arise because a complete correspondence between original language and target language is almost impossible. Audiences are often aware that something of the essence of media content is lost in translation (Rohn, 2010, p. 81). Translation of foreign television text is subjective. A successful translation is based on a translator's very good awareness and knowledge of the foreign socio-cultural context.

The quality of a translation has great impact on watching foreign programs. If a translation was "accurate", interviewees could "really understand foreign television programs". If translation had "some misunderstanding or misrepresentation" or it was "not according to the original meanings", they could "not really understand the foreign television programs" (Interview SH_3). Zhang from Guangzhou and Zhou from Xi'An had a similar opinion about the quality of translations:

> "If a translation is good, I would like to watch it. However, if translation is bad, even good television programs will be not interesting." (Interview GZ_3)
>
> "If a translation is good and successful, then I will have a good feeling about this television program. However, if differences between the translation and the original meaning are obvious, then I will not enjoy the program." (Interview XA_9)

The quality of a translation in the opinion of the interviewees included two aspects: accuracy of the translation of meaning, on the one hand, and accuracy of the translation of humor, allusion and emotion, on the other.

*(1) Accuracy of translation of meaning*

Both forms of linguistic translation – dubbing or subtitling – naturally modify the text to some degree (Collins, 1998, p. 134). Technical considerations regarding the length of utterances and needs for audiovisual dubbing are often a problem which leads to cutting translations short and omitting parts of text (Chaume, 2008, p. 134). There is also an ever-present danger of changing the meanings of the text or there being a difficulty of translating certain words or terms that might be unavailable in the destination culture. Finally, some errors in translation are almost inevitable and the text as presented in the destination country might remain only slightly unchanged. (Cohen, 1992, p. 25)

Firstly, some interviewees could not perceive any difference between the original language and translation. Generally, they found that "subtitles of foreign movies in cinema corresponded with the original meanings relatively well" (Interview BJ_6) and they "cannot feel any unsuitable translations" (Interview SH_5).

Wang from Shanghai watched foreign television programs through *Xun Lei* which is a Chinese software package for downloading television programs from the internet. She thought that the translations of foreign programs on *Xun Lei* were "very good and accurate", because "translators are very familiar with foreign cultures" and they not only translate "literal meanings of original language", but also "emotion and meanings at a deep level of speakers". (Interview SH_3)

Secondly, some others found differences to be very "obvious" and "big". The translation was sometimes "strange", "wrong", "not correct", "not very accurate", "arbitrary", "confusing", and even "not understandable". However, most viewers could "accept" and "understand" most of translation.

Wang explained further that the translation of some foreign programs, which were broadcast through television (e.g., some movie channels of Chinese television stations), was "not very good". Not only the "meanings of the original language", but also the "emotions" were translated "very poorly". She could feel that it was quite obviously a translation. (Interview SH_3) Wu (a manager from Beijing) said:

> "When you watch the dubbed foreign programs, you have the feeling like you are 'reading' the feelings of other people. It is not original anymore. The dubbing is something that has been integrated into another unfamiliar culture. Therefore, it is not a good way to understand the original meanings through dubbing. Some foreign words cannot be translated into Chinese." (Interview BJ_7)

*(2) Translation of humor*

Puns, word plays, jokes, and humor vary considerably from culture to culture and in general often pose insurmountable problems in translation (Chiaro, 2010, p. xvi). If they cannot be reproduced in any way, the translator might attempt a word play in a different place in the target language. However, this is a controversial method, because the original author put this word play in a certain place for a reason. Sometimes, rather than risking such chancy operations, a translator will be better off leaving the humor out altogether. If the purpose of the source-language text is to create laughter which cannot be served, a word-by-word translation, even with an explanation, seems superfluous. (Gonzalez, 1989, p. 488)

Among these word plays humor differences between the original language and the translation were perceived by participants the most. One interviewee could not understand humor in foreign programs, such as the American comedy show *Friends*. Ding and Peng from Beijing said:

> "Humor in foreign programs is something different to our way of thinking. Sometimes I cannot react to humor like a foreigner would, so don't find it funny." (Interview BJ_2)
> "Sometimes I cannot understand their humor. Perhaps they might laugh, but I think it is unusual and unfunny." (Interview BJ_4)

## 10.3.2 Values

A value is a criterion for selecting objects that people should strive for, a normative standard for evaluating objects as well as approving opinions. Values determine how people see the world and behave in it (Grimm, 2003, p. 21). Generally, values are understood as general notions about what is good and what is bad, what is desirable and what is undesirable. (Berger, 1992, p. 16)

Values as a cultural-difference dimension in television programs hide behind the manifest content of pictures, sounds and characters. They are an abstract but important element of TV programs (Wood, 2010, p. 304). Implicit in this view is firstly that producers create television programs that reflect their own basic values and, secondly, that audiences select from a wide variety of television programs those that are most consistent with their worldview (Berger, 2011, p. 123). What values are espoused and what values are attacked, downplayed, or neglected in television programs vary between cultures (Samovar, 2011, p. 59). Values in television programs are perceived by viewers based on their everyday life and understood in a cultural context. They can use value principles that they have seen in television programs to judge the world and to examine their own value principles. (Grimm, 2003, p. 19)

The results of interviews indicated, on one hand, that interviewees still have a strong awareness of Chinese values (e.g., the Confucian set of beliefs) in their mind. A consequence of this attitude was a high priority given to Chinese values, but a low priority to values in foreign programs. On the other hand, results also demonstrated that Western media did indeed have some effects of Westernizing interviewees' values. Western values had crept into China and influenced some aspects of Chinese values.

Based on numerous statements by the interviewees, values were conceptualized on three levels:

- Individual level – "individualism", "religion", and "morality";
- Family level – "family centralism";
- Social level – "liberty/democracy/equality", "social status/competition", and "patriotism"

Pan from Guangzhou used the words "freedom", "pride of their countries", "independent", "individual", "social responsibility", and "united" to summarize how the typical values in foreign television programs looked:

> "They like freedom. They are proud of their countries. They are independent and have a strong awareness of individuals. They also have a strong social responsibility and a strong ability to be united with each other." (Interview GZ_5)

- **Individualism**

Hofstede defined this term as follows: "Individualism stands for a society in which the ties between individuals are loose; everyone is expected to look after himself or herself and his or her immediate family only," and "collectivism stands for a society in which people from birth onwards are integrated into strong, cohesive in-groups, which throughout people's lifetime continue to protect them in change for unquestioning loyalty." (Hofstede, 1991, pp. 260-261)

Individualism, on the one side, versus its opposite collectivism is the degree to which individuals are supposed to look after themselves or remain integrated in groups, usually around family (Andersen, 2003, p. 77). Individualism is characterized by individual autonomy and a preference for individual interests and goals over group interests and goals (Brewer, 2010, p. 312). Thus, in individualist countries, people tend to think more about their own personal success, whereas in collectivist countries loyalty to others is very important. (Niemeier, 1997, p. 283)

Traditionally, Chinese are integrally connected with other people and concerned more about the groups they belong to, (Broadbeck, 2007, p. 1032) such as they and their parents, children, and relatives as a family group; they and their colleagues as a work group; and they and their friends as a friend group. Chinese consider the health and security of their family members, as well as the interests and welfare of their companies as being more important than their own. Even having dinner in China is a collective experience. Chinese are used to having dinner with some people together sitting around a round table. They share all of the dishes with each other.

Although it is difficult to say in an era of globalization whether Chinese values are being eroded by Western values, they are clearly not individualistic. The "self" is not a featured concept in traditional and current Confucian teaching in China. Western media, however, unknowingly transmit messages that promote independence, privacy, equality, and freedom of individual. (Chaffee, 1992, p. 45)

In the interviews some stated that values did "not have an essential difference" (Interview BJ_7). Some others found value difference to be "large" and "obvious", because "Western countries advocate democratic and individualism" whereas "Chinese traditional culture advocates collectivism" (Interview BJ_7).

Individualism in the minds of the interviewees meant that foreigners in TV shows seemed more "self-centered". They paid a lot of attention to the feelings of themselves and "individual interests". Collectivism meant a "stronger awareness of family, parents and children". They thought the Chinese paid more attention to the "interests of country and collective interests", sometimes even "set aside their own interests to protect the collective interests" (Interview XA_9). Wang from Shanghai explained:

> "The foreign movies, such as American movies, stress individual heroism; in other words, their individualism is relatively strong. Domestic Chinese movies or television series emphasize the traditional orthodox values or mainstream values. They pay little attention to individuals." (Interview SH_3)

- **Religion**

In the interviews religious difference was the most obvious and the biggest cultural difference in foreign TV programs of all cultural-difference types. This result is related to religious tradition in China. Most Chinese in China are atheist and do not hold religious beliefs. (Yang, 2012, p. 3) The traditional Chinese religions, such as Buddhism and Taoism, are gradually losing their significance, because the young generations in China learn a lot about atheist beliefs like those of Mao and Marx in school or college (Knight, 2005, p. 26). Religion is also a sensitive issue in daily life and people rarely talk about it. Thus, they have little chance of learning anything about religion.

The majority of interviewees had tried to establish a clear boundary between what they considered as "real" (a positive term) and "not real" (a term related to religion). The statement used most to describe about religious content in foreign television programs was "religion is far removed from my real life". In this case it could be determined that religion difference had reached a very high level.

In general, Chinese viewers thought that religion was an important and necessary part of life in foreign countries, because in foreign programs they could often see foreigners going to church and taking part in religious ceremonies. However, Chinese television programs had hardly any religious content, because most Chinese are atheist and only a small group of them hold religious beliefs.

In this study none of the 36 interviewees had religion beliefs. Some said they could "understand" religion" and they thought "it is very good to have religious beliefs", but they did not "worship or believe in it" (Interview GZ_3). Some said that they "respect" religion beliefs, but they did "not take part in any religion activities" (Interview GZ_4). Some others explained they "do not believe in religion" and were "not interested in religion", because they "believe in science" (Interview NJ_3). For others, "religion does not have any impact" on them and they "do not have any feelings about religion". Hu from Nanjing explained why he did not believe religion, but he could understand why a lot of people did:

> "In the hearts of everyone there is a kind of belief and they want to rely on someone else. However, if someone else cannot realize their belief, they will rely on God for this. Everybody can understand this." (Interview NJ_3)

Although none of 36 interviewees had religious beliefs, their opinions about religion in foreign television programs were quite controversial:

(1) Some participants did not like or could not accept religious content in foreign TV programs because they thought religion was "false" (Interview NJ_6) and "far removed from my life" (Interview XA_7). Someone said "religion ceremony looks a little strange" for them (Interview SH_4).

(2) Some others could understand and accept religious content in foreign programs "partially", but "not totally". Whether or not they accepted these pro-

grams was related to how much religious content a program contained. If the "whole" program was concerned with religion, they could not accept it; if "there is just a little content about religion", they could accept it (Interview NJ_3).

(3) Just a few participants could totally understand and accept religious content in foreign TV programs. How was it they did not believe in religion, but were able to accept it totally? The reasons were varied:

- Their minds were "very open". They believed that "cultures around the world are diverse". Thus, "each nation has its own special characteristics" (Interview XA_7) and "own special traditions and rituals" (Interview XA_3);
- In China "many people believe in religion" (Interview XA_2) and "actually, China also advocates religious freedom" (Interview XA_3).
- Religion was a "kind of ideological domination" in foreign countries. "Socialism" was also a kind of ideology in China. Some interviewees believed that they were essentially "the same", although "the way they are performed is different" (Interview XA_9).

(4) Only a few participants liked religious content. Wang from Shanghai "appreciated" religion (e.g., "Islam and Christianity") in foreign television programs. She believed that both in China and foreign countries "religion is a kind of culture on a very deep level". When she watched programs with religious content, she had a feeling of "shock and baptism", because "the religion ceremonies are very serious and religion can change people in their very soul". The most important reason for why she appreciated religion was that "religious beliefs are good for individuals and for the whole nation". (Interview SH_3) She explained:

> "Most people who have their own religious beliefs will not lose themselves in life and society, because they behave according to the rules of their religion. Now in China many Chinese are rich, but they do not have religious beliefs. Their beliefs only concern money and private interests. Such beliefs are very superficial. Therefore, it is easy for them to lose themselves. The whole of China lacks religion beliefs. Thus, the entire nation sometimes seems very impetuous. [...] If the individuals in a nation have common religion beliefs, they will feel a closer bond with each other. If they do not and they just associate with each other for their own private interests, the nation will not be very strong. This is my personal opinion." (Interview SH_3)

- **Morality**

The statements by the interviewees indicated that the definitions and judgments of morality and immorality in foreign television programs and in China were different. Moral standards in China "come primarily from Confucianism" and are strongly influenced by traditional Chinese culture, while moral standards in

Western countries "come from their laws" (Interview SH_3). Generally, most interviewees found that the moral standards in foreign TV shows were "more open" and "freer" than Chinese moral standards. Wang said:

> "Although China has been opened up to the outside world for many years, the morality of the Chinese is still deeply influenced by traditional Chinese culture, which has developed for several thousand years." (Interview SH_3)

(1) *Not accept*: Some interviewees did not like or could not accept certain content in foreign TV programs because it was "immoral" based on their own moral standards, for example, the use of sexuality in television commercials or content about homosexuals. They thought that some foreign programs tended to be too risqué and overtly sexual, because the "relationships between men and women as depicted in some foreign television programs" seemed too "crazy" as "China is relatively conservative" (Interview GZ_3). Ding from Beijing explained:

> "The relationships between women and men, such as in bars, are much more open than in China. I cannot say that the Chinese are more conservative than they are. The difference is that such relationships between women and men can been seen in foreign programs. Chinese television programs do not show anything like that." (Interview BJ_2)

(2) *Accept*: Others thought there was no significant cultural difference in terms of morality and they could "gradually" accept it as they get to "know more about them" (Interview BJ_2). Xu from Xi'An did not think that moral standards for adjudging what should be prohibited content for television bore any cultural differences, because foreign programs "have their own characteristics and styles". It was just a difference in television styles – not a moral difference. (Interview XA_8)

(3) *Preference*: Some other interviewees found the morality in foreign television programs to be "better" (Interview XA_7) than in Chinese shows. Shi (a lawyer from Guangzhou) accepted the moral standards in foreign programs as "very good". He said that when he was a child his teachers and parents told him that foreigner's moral standards were "not good". However, he did not think this was true. Based on his experience working as a lawyer in China he said:

> "In our country now there are really a lot of immoral things. This problem is related to moral education. If traditional moral standards still played a role in current society, there would not be so many problems. Developed countries also have some immoral things, but they are not as widespread as they are in China. [...] Perhaps the developed countries have also gone through this stage of developing the morals of its citizens. Now we are in the stage of developing our morals." (Interview GZ_6)

- **Family**

The traditional Chinese family values are based around family centralism, which views the family as the center of a person's universe and the basic building block of society (Bell, 2010, p. 11). Indeed, in traditional Chinese culture there was no concept of the individual: one's identity was completely defined through one's relationships with family members, and one's life was governed by a complex web of obligations within the family network (Kleinman, 2011, p. 3). For example, in China parents must take good care of their young children. When the parents are old, it is then the responsibility of children to look after their parents. If children do not do this, they are considered as not honoring and not respecting their parents. (Sheng, 2005, pp. 223-225)

Some interviewees believed that the family values in foreign programs were "far removed" from Chinese traditional family values (Interview XA_2). The awareness of the family in China was "stronger" than that shown in foreign television programs (Interview BJ_6). For example, Zhang (a student from Shanghai) said, "the relationship between children and parents seems cold" in foreign programs. (Interview SH_4) Although the cultural difference of family values was great, they could also "understand" and "accept" it (Interview SH_5).

In the opinion of some interviewees, Asian countries (e.g., China, Japan and Korea) shared similar "attitudes towards the family" which were "relatively conservative" and "quite different" compared to the attitudes towards the family in Western countries (Interview GZ_1). Zhang (an international trader from Guangzhou) believed that the tradition of family values in Korean was better inherited than in China. He explained:

> "In Korean movies and television series we can see that children comply with their parents' wishes and respect their elders. It is a traditional Asian culture, which has been very well preserved in Korea. However, in China now, many married Chinese do not comply with their parents' wishes or respect them. Sometimes they even shout at their parents." (Interview GZ_3)

- **Liberty/Democracy/Equality**

Many interviewees could see in foreign television programs that their "values", "ways of thinking" were "liberal", "democratic" and "equal".

Some interviewees saw from foreign television that people were "very liberal when they do anything" and "they are very brave, much more than one could imagine the Chinese being" (Interview BJ_1). In contrast, Chinese television programs were "usually relatively traditional" and concerned little with liberal thinking.

Equality means treating people impartially and without favoritism as well as there being justice for everyone regardless of social or economic circumstances. (Hall, 1990, p. 150) For example, “they display little race discrimination” (Interview SH_4), and “they respect different people of different races” (Interview SH_5).

Some participants found the liberty, democracy and equality, depicted in foreign television programs, to be very “nice” and they liked it. Han from Xi’An believed that Chinese traditional values had gradually changed. Now in China people “also advocate democratic and liberality” (Interview XA_3).

- **Social Status/Competition**

Some participants perceived that people in foreign television programs had a “strong awareness” of “authority”, “power”, and “competition”. Zhang (an international trader from Guangzhou) expressed his opinions about social status and competition in foreign television programs, especially in programs about elections:

> “In some foreign television programs people pay a lot of attention to and pursue money and social status, such as movies and television programs about elections. If one person wants to run in an election, everything he has done in the previous 10 years will be scrutinized in order to show his true character. Therefore, they pay close attention to social status.” (Interview GZ_3)

This strong awareness of social status and competition is a tradition of some Western countries. For example, in the United States, most people admire hard work and success in business and life. They are competitive and do not feel hesitant about enjoying it. Their strong drive for recognition is evidenced in their stance, dress, posture, attitude, voice level, and possessions. Their heroes are public figures who frequently appear on television: outstanding athletes, Hollywood actors, television stars, even business leaders. (Hall, 1990, p. 149) Thus, Chinese viewers could obviously also perceive the strong awareness of social status and competition in American television programs.

- **Patriotism**

Some interviewees could sense the different ways of expressing feelings of state pride and patriotism in foreign and Chinese television programs. In general, the Chinese express these patriotic feelings more directly, such as by shouting “I love our country”, while people in foreign programs express their patriotic feelings rather in a more implicit and indirect way, such as using the national flag or some other symbols to show their pride in their country. Zhang from Guangzhou said:

> "In television programs in every country people seem to love their countries, no matter whether it be Korea, Japan or America. I often see American flags in many American television programs. This is more or less communicating that American citizens are proud to be American and they love their country." (Interview GZ_3)

## 10.3.3 Way of life

Way of life is created through the long-term cultural development of everyday reality, related events and objects, as well as individual life experiences. A certain way of life symbolizes identity as well as a feeling of membership of a culture, which clearly distinguishes it from other ways of life (Bachmair, 1996, p. 278).

Way of life as a cultural difference in television programs can be recognized in numerous habits, customs and rituals, as well as living environment. *Customs* and *rituals* can be understood as standard cultural acts. Some interviewees considered that the following aspects in foreign programs were different compared to those in China:

- Habits (e.g., greetings and cooking);
- Customs and rituals (e.g., weddings and traditional celebrations);
- Living environment (e.g., landscape and architectural style).

- **Habits, Customs, Rituals**

Many interviewees believed that the cultural difference in regard to habits, style of dress, and rituals was still great. Zhang (a Scholar from Xi'An) thought that the cultural differences in way of life between China and foreign countries had been very large before the "*New Life Movement*" (XinShengHuo YunDong) in the 1930s in China. Since then, the differences "become smaller", although "currently there are still some differences" (Interview XA_10).

These differences in the words of the interviewees included "way of greeting", "cooking tradition", "food culture", "driving habits", "activities of people", "clothing", and so on. Meng from Xi'An said, "The Chinese shake hands when they meet someone. However, I see people in foreign television programs who do not do this." (Interview XA_4) Guo from Guangzhou regarded the distance between people when they are talking to each other as a cultural difference of habits. He said:

> "The Chinese like to talk with other people in close bodily proximity. In some countries, people stand a meter away from each other when talking. Therefore, they are not comfortable talking with each other if there is less than a meter between them, unlike the Chinese." (Interview GZ_4)

The most stated differences of way of life concerned "party" and "wedding":

*Party*: Interviewees said that they could "often" see parties in foreign television programs, whereas parties were "rarely held" in China. Feng (a student from Bejing) saw that foreigners hold birthday parties in bars in foreign movies and television series, whereas on her birthday she "just invites a few friends over and they cook something together" (Interview BJ_6).

A lot of viewers thought parties in foreign countries were very good, because they looked "more casual, active, and open" than in China. Furthermore, their "parties are a good chance to allow people to get together, communicate with each other, and relax." (Interview GZ_3) Another interviewee did not seem to like parties in foreign programs: "I think their awareness of entertainment is very strong. Sometimes they are crazy. I do not like that very much." (Interview SH_5) Zhao (a worker from Xi'An) said:

> "The style of their parties is different as ours. I feel that foreigners in parties are very open and liberal. We belong to a more conservative culture. [...] Sometimes parties in foreign movies are actually not acceptable for most people in China. However, I think the foreign movie makers surely have their own agenda and ideals for expressing something through these parties." (Interview XA_8)

*Wedding*: Opinions about weddings were also controversial. One interviewee liked the Chinese traditional wedding ceremony "better" than Western weddings. In contrast, more people liked and accepted Western wedding more than Chinese ones, and they wanted to their wedding to be just like what they had seen in foreign television programs (Interview GZ_3). Yang from Guangzhou did not like Chinese weddings, because "some weddings in China are too big and overly lavish, especially those of some rich people, who invite too many people to their wedding" (Interview GZ_2).

Some said that Western weddings had become an aspect of the new life style in China. People, especially in big cities, also "go to churches to hold their weddings just like the wedding ceremonies in foreign countries". However, Dr. Zhu from Xi'An explained that the reason for doing this were different: "foreigners holding their weddings in church means that weddings for them are very sacred", while "the Chinese in China holding their weddings in church is only a way of saying to other people that they are getting married" (Interview XA_7).

He stated that the Chinese did not really know the religion meaning of a wedding being held in church. Wang from Shanghai was already used to seeing Western weddings in China:

> "Now in many modern Chinese movies and television series we can see that Chinese weddings look like the weddings in Western countries. [...] They learn from Western developed countries. [...] When I watch Western weddings in foreign television programs, I am very familiar with them, because such things are often happening around us." (Interview SH_3)

- **Living environment**

The living environment, such as landscape and architectural style, in foreign television programs sometimes seemed "exotic" to Chinese viewers, but they did not have any "strange feelings" about this like they used to. In contrast, some places, e.g., Western cafes, restaurants, and pizzerias, are also very popular in China now and a familiar part of everyday life in China. Like Westerners, many Chinese also go to *Starbucks*, *Pizza Hut*, and *McDonalds* for lunch to or meet someone. (Interview BJ_5) Thus, they did not see any great difference in this area between foreign countries and China.

### 10.3.4 Education Method

Different cultures have different methods and patterns of education. For example, parents' overindulgence in their children is much more common in China than elsewhere (Lim, 2006, p. 188). Children in some Asian cultures develop less of an individual self and more of a family self than in some Western countries. This makes them more sensitive and responsive to their surroundings. People in some Western developed countries, on the other hand, are puzzled by people for whom the self is not just the individual. (Saxon, 2006, p. 59) The notion that the self is not really a person but in fact the nexus of a web of relationships – family, friends, and peers – all of which entail obligations, is difficult for people in Western countries to accept. (Nolan, 1999, p. 9)

Most interviewees said that the education methods, which they had seen in foreign programs, were more liberal and open than in China. A consequence of this was that people in foreign programs seemed more independent, for example, they could find solutions to problems or make decisions by themselves. In contrast, the Chinese were more dependent on each other and have to show obedience to parents, elders, and superiors. Therefore, they have little chance of finding solutions or making decisions by themselves.

- **Freedom/independence in education**

Because family is an important concept to the Chinese, children have customarily been viewed as the center of the family. Consequently, there has long been a Chinese tendency to dote on children. Parents and grandparents are extremely protective. A consequence of this is that children's awareness of independence is very weak. For example, they do not need to do their housework and do not have to work to pay their school fees, because their parents take care of it.

Some interviewees, who were parents, said that they could see obviously a difference in regard to education methods. A mother of a child from Nanjing said, "Foreign parents pay less attention and take less care of their children than Chinese parents." (Interview NJ_2) For example, "children in foreign countries can work once they are 18." (Interview NJ_6) Yang from Guangzhou believed that the education methods she had seen in foreign programs were obviously "more open" and "better" than in China, because foreign parents gave their children more freedom. She said:

> "In China, parents consider their children as treasures and as little emperors in a family. They always take very good care of them and are afraid that something might happen to them. [...] Foreigners also take care of the security of their children, but they do not protect children as much as Chinese parents do." (Interview GZ_2)

Education methods in China cover not only children, but also adults, such as those working in a team. Zhang (an international trader from Guangzhou) watched a foreign movie named *The Greatest Game Ever Played*, which told the story of a team of golfers. He thought this American movie had very good educational significance. He described it as follows:

> "The golf trainer's name is Harry Vardon. [...] He respects his team members. He does not give them direct answers to questions. However, he gives them suggestions and guidance that enables them find answers on their own. In contrast, parents in China tell their children directly what children should do and they always make decisions for them. Therefore, Chinese children are used to obeying their parents' orders." (Interview GZ_3)

- **Imagination/creativity**

In traditional Chinese education the imagination and creativity of children is not considered important. In contrast, repeating facts out loud from textbooks is considered important. (Jolley, 2010, p. 275) This is often called "YingShi JiaoYu" (exam-oriented education) (Zhou, 2008, p. 55). Many Chinese children study very hard in school and college primarily to attain very good marks in exams, not to develop their imagination and creativity. Children are always taught to recognize that everything in the textbook is correct and to obey their parents and teachers (Wong, 1992, p. 38). So children's imagination and creativity gradually diminishes.

In China today, many Chinese parents and teachers are beginning to pay more attention to imagination and creativity during children's education. (Bao, 2004, p. 25) However, compared with some Western developed countries imagination and creativity in Chinese education is still at the beginning stage.

- **Equality between young and old**

Chinese culture is a typical culture with a high power distance. Power distance means to what extent power in organization and family is distributed equally. (Chapel, 1998, p. 177) High power distance indicates that the less powerful members of the organization or family must respect people with more power. For example, in Chinese families young people should respect old people and obey their instructions.

Shi (a lawyer from Guangzhou) thought the education methods in foreign television programs seemed "very good", because "their parents communicate with and talk to their young children in an equal way just like friends". He would like to treat his child in "this friendly and equal way" in the future. (Interview GZ_6) Zhang from Guangzhou shared the same opinion. He would also like to educate his child using the methods he had seen in foreign programs. He would not allow his parents to "use their traditional education methods to educate" his child like they had educated him. He related the following story from his childhood:

> "If I wanted to get some pocket money from my parents, I had to do some chores for them, such as sweep the floor or wash the dishes. If I did not do any housework, I would not get any pocket money." (Interview GZ_3)

- **Demand on Children**

Zheng (a schoolgirl from Nanjing) was candid in her assertion that Chinese parents' demands on children are too high. Thus, she found the education methods in foreign countries were "very good". She said:

> "I have seen in foreign television programs that the demands of parents on their children are lower than in China. Chinese parents impose all of their hopes on their child. I think this is too stressful for children in China." (Interview NJ_6)

One important reason for such high demands on Chinese children is China's one-child policy, which has been in place since the 1980s. A consequence of this is that the sole child then represents the hopes of a family, because the parents of this new generation rarely had the opportunity to study in college. Due to economic problems during their childhood, they had to work hard and save. Therefore, Chinese parents do not want their children to "eat bitterness" like they did (Hua, 2011, p. 190). Instead, they have high demands on their children's education and want their children to get good jobs in the future.

A Chinese documentary *Qing TouWo YiPiao* (Please Cast Me A Vote!), which tells a true story about an election for class leader in a Chinese primary school, is a very good illustration of the high demands of Chinese parents and the competition awareness of Chinese children.

## 10.4 Catalogue of Cultural-Difference Dimensions

On the basis of numerous statements about cultural differences made by the 36 Chinese study participants in five sample cities, the author developed a catalogue of cultural-difference dimensions in foreign television programs. A total of 10 sets of dimensions/42 dimensions were summarized in this catalogue. (See Table 13)

All cultural-difference dimensions are divided into two groups:

(1) "Television-genre-*related* dimensions" are related to certain content and formats of TV programs (incl. movies & TV series, sports programs, documentaries, entertainment shows, news programs, and music programs);

(2) "Television-genre-*unrelated* dimensions" are not related to certain content and formats of TV programs (incl. language, values, ways of life, and education methods).

Four principals were considered by summarizing the cultural-difference catalogue:

(1) The selected items should be important and typical for the interpretation of cultural difference in foreign television programs;

(2) Each item should be clearly distinguished from the others, meaning that redundant items with similar meanings should be omitted;

(3) The items should be interpreted using the words of interviewees as much as possible, thus in the later survey part of the research they will be able to understand these items better;

(4) The items should be interpreted as concisely as possible, because complicated interpretations could cause confusion for Chinese participants who do not require much information in a high cultural context.

In this catalogue all dimensions are divided according to three cultural levels:

(1) *Level A is the surface level*: Here everything is made explicitly clear and is highly visible. This describes the manifestations of culture that can be observed using the senses. Television viewers should generally find the fewest intercultural misunderstandings at this level.

(2) *Level B is the middle level*: Things here are at least partially visible. It may appear that a culture's underlying beliefs are being revealed, but in reality it represents how that culture presents itself. It includes norms, rules, traditions, rituals, and customs of a culture.

(3) *Level C is the deep level*: It is the essence of culture and is largely invisible and unconscious. It includes the basic norms, core values and beliefs that shape a culture's worldview and influence cultural identity.

However, cultural dimensions at the surface level in television programs can be modified over time, such as the appearance of characters. However, uncovering and deciphering cultural dimensions at a deep level is difficult, because they are not accessible at the surface attitudinal level and viewers need a long time to understand them step by step, such as, for example, religion and worldview in foreign television programs.

*Table 13: Catalogue of Cultural-Difference Dimensions*

| Part One | Genre-Related Cultural Difference | Level |
|---|---|---|
| Set 01 | Movie & TV series | |
| 01 - 1 | Appearance of character | A |
| 01 - 2 | Personality of character | A |
| 01 - 3 | Music style | A |
| 01 - 4 | Theme of drama plot | A |
| 01 - 5 | Proximity to daily life of drama plot | A |
| 01 - 6 | Fantasy of drama plot | A |
| Set 02 | Sports program | |
| 02 - 1 | Moderation/commentary | B |
| 02 - 2 | Sports culture | B |
| 02 - 3 | Commercialization | B |
| 02 - 4 | Sports types | A |
| Set 03 | Documentary | |

| | | |
|---|---|---|
| 03 - 1 | Objectivity vs. subjectivity | B |
| 03 - 2 | Explicit vs. implicit | B |
| 03 - 3 | Starting point/perspective | B |
| 03 - 4 | Theme | A |
| Set 04 | Entertainment show | |
| 04 - 1 | Speaking style of presenter | A |
| 04 - 2 | Content of entertainment | A |
| Set 05 | News | |
| 05 - 1 | Appearence of presenter | A |
| 05 - 2 | Personality of presenter | A |
| 05 - 3 | Objectivity vs. subjectivity | B |
| 05 - 4 | Starting point/perspective | B |
| 05 - 5 | Diversity of news theme | B |
| 05 - 6 | Sensitivity of news theme | B |
| Set 06 | Music program | |
| 06 - 1 | Music style | A |
| 06 - 2 | Music quality | A |
| **Part Two** | **Genre-Unrelated Cultural Difference** | **Level** |
| Set 07 | Language | |
| 07 - 1 | Voice of dubbing | A |
| 07 - 2 | Emotion of dubbing | B |
| 07 - 3 | Coordination of dubbling | B |
| 07 - 4 | Accuracy of translation | B |
| 07 - 5 | Humor in translation | B |
| Set 08 | Value | |
| 08 - 1 | Individualism | C |
| 08 - 2 | Religion | C |
| 08 - 3 | Morality | C |
| 08 - 4 | Family | C |

| | | |
|---|---|---|
| 08 - 5 | Liberty/democracy/equality | C |
| 08 - 6 | Social status/competition | C |
| 08 - 7 | Patriotism | C |
| Set 09 | Way of life | |
| 09 - 1 | Habits/customs/rituals | B |
| 09 - 2 | Living environment | A |
| Set 10 | Education method | |
| 10 - 1 | Freedom/independence | B |
| 10 - 2 | Imagination/creativity | B |
| 10 - 3 | Equality between young and old | B |
| 10 - 4 | Demand on children | B |

Source: Own illustration.

# 11 Results II: Chinese Television Viewer Types

The second part of the interview results concern television viewer types. The results indicated that Chinese viewers could still more or less perceive some kinds of cultural differences, but some differencсs were thought to be surmountable and some viewers downplayed the significance of this as merely minimization.

The cultural adaptation process for viewers consisted of three stages:

(1) *Culture shock*: In this first stage, viewers are continually surprises by what they encounter when they watch foreign programs for the first time. However, this is a normal and near-universal response to radical changes in a cultural context witnessed when watching television. They have lots of difficulties understanding foreign programs, because cultural differences are perceived as being very obvious and large. It is not possible for them to immediately accept such differences. Viewers, who like culture difference, find everything in foreign programs to be charming, exotic, and exciting. Viewers, who reject cultural difference, will change channels and watch somewhat more to their liking. They can ignore these foreign programs, pretend they do not exist, and attempt to erect barriers in order to isolate their own culture from foreign ones.

(2) *Cultural adjustment*: In the second stage, viewers' opinions begin to shift slightly. They embark on a learning process to try to minimize the differences they have perceived and attempt to find some sort of common ground they can relate to. Once viewers start to try to understand cultural differences, they can then spend decades learning to become comfortable watching the strange and unfamiliar things they see in foreign programs, which do not correspond to their own familiar culture.

Being receptive and keeping an open mind is probably the most helpful attitude for facilitating cultural adjustment. If they are adaptable, tolerant, patient, and are also curious and ready to learn new cultures, then they are off to a good start. Then, the search for common ground is successful and there is a high probability that they will gradually come to accept foreign programs. Thus, they already have some sort of cross-cultural experience to deal with new and unfamiliar cultures. Much of what they encounter is probably somewhat familiar already. However, there are some important differences too, which can still be perceived.

In contrast, if viewers are impatient, highly strung or fearful of change, the probability of successfully finding common ground will be very low. In this situation they go back to focusing on distinctions – seeing themselves as fundamentally unlike "these people" in foreign programming. Cultural differences appear quaint and exotic to them. Finally, if they have accumulated frustrations and setbacks in their attempts to understand cultural difference, they will start to exaggerate these differences, and consider them as insurmountable as well as negative obstacles.

This cultural adaptation process is a learning experience, as viewers make the transition from one cultural frame of reference to two or more. The benefits of a successful cross-cultural adjustment are considerable: they learn a great deal about another way of life and gain a new insight into their own culture.

(3) *Cultural adaptation*: In this stage, viewers cannot perceive any cultural difference, but accept them totally. They could become part of the culture that is presented in foreign programming. It does not mean they will give up their own culture and its values; their adjustment will remained anchored to their own culture during this process.

## 11.1 Five Television Viewer Types

The results from the personal interviews indicated that all perceived cultural differences varied between "large" and "small", while their degree of acceptance varied from "not acceptable" to "completely acceptable". According to the degree of perception and acceptance, the author divided all interviewees into five groups: escapees, learners, adherents, adjusters, and neutralists: (See Table 14)

(1) *Escapees*: Viewers who perceived cultural difference to be large and did not accept it. Their attitude toward cultural differences was to avoid them, because they knew little about foreign cultures and their minds were not open to learning something new about other cultures.

(2) *Learners*: Viewers who could also perceive large cultural differences like escapees, but they are also happy to accepting them. Their minds were open to learning foreign cultures through watching foreign television. They would like to participate in the process of globalization and part of it. Thus, they did not have much difficulty understanding cultural difference. Through learning they felt it less and less.

(3) *Adherents*: Viewers who perceived cultural difference to be small, but they still adhered to traditional Chinese culture because their opinions were strongly influenced by their traditional culture. They were not ready to accept much cultural difference.

(4) *Adjusters*: Viewers who believed that cultural differences were very small and accepted almost all of them. They were very interested in cultural differences and glad to see differences in foreign television programs, because they wanted to know more about other cultures.

(5) *Neutralists*: Viewers who are somewhere in the middle of the other four viewer groups. They neither paid attention to cultural difference, nor wanted to adapt it. They believed that cultural differences were neither large nor small. They could neither accept them nor not accept them.

*Table 14: Types of Chinese Television Viewers*

| | Viewer Type | Degree of Perception | Degree of Acceptance | N |
|---|---|---|---|---|
| A | Escapee | High | Low | 9 |
| B | Learner | High | High | 13 |
| C | Adherent | Low | Low | 0 |
| D | Adjuster | Low | High | 10 |
| E | Neutralist | Middle | Middle | 2 |
| | | | Total | 34 |

Note: A classification of two interviewees is not possible.
Source: Own illustration.

In the following, five viewer types are described in detail together with an introduction of some cases. The interview results clearly demonstrated that each participant had their own particular attitude towards cultural differences. However, it is important to point out that using five viewer types is an ideal way to categorize Chinese television viewers. In some cases, the statements of participants revealed controversial attitudes. Furthermore, their attitudes might vary when reacting to various cultural-difference types.

### 11.1.1 Escapees: High Perception with Low Acceptance

- **"It is more realistic to watch some Chinese programs."**

Some participants (9 out of 36) as escapees reacted to cultural difference in foreign television with a low acceptance and held a deep to cultural difference. This meant they were unwilling or unable to confront the adaptive demands of viewing foreign culture from an insider's perspective.

Escapees are, for example, middle-age or old people in China born before the 1970s. They grow up in a relatively conservative environment, because at that time China had not been opened up to the outside world. Before 1980, there were no foreign television programs, foreign newspapers and magazines, and no internet in China. Thus, there was little opportunity to learn anything about foreign cultures.

*Case 1: Yin, a 19-year-old student from Xi'An (bachelor degree)*

He is a typical escapee. He believed that "foreign television programs definitely have some cultural difference compared to China". He said that he did "not like foreign programs very much", because he lived in China and was "familiar with Chinese culture"; moreover, he said he was "not familiar with foreign culture" (Interview XA_1).

*Case 2: Zhao, a 43-year-old worker from Nanjing (middle school)*

She watched very few foreign television programs, referring to watch Chinese programs. She said that she did "not like foreign television programs" and she was "not used" to watching them. If she had free time at home, she only watched some news about foreign countries through CCTV-4 (international Chanel) and CCTV-12 (social Chanel). She explained that "we are influenced primarily by traditional Chinese culture and habits. The people my age like Chinese tradition." (Interview NJ_8)

*Case 3: Jiang, a 41-year-old man from Nanjing (bachelor degree)*

He was the father of a 4-year-old child. He had prejudiced attitudes towards foreign education programs for children. He did not allow his child to watch foreign children programs and only let him watch some domestic cartoons. He explained that "their way of thinking has some differences compared to ours" and "for a child's education, domestic cartoons are better", because "the content in foreign cartoons is not related to our lives in China" and "their impact on children is not very good". (Interview NJ_7)

*Case 4: Li, a 38-year-old man from Beijing (middle school)*

He did not watch any foreign television programs, because he could perceive cultural differences very easily and "knows little about foreign television programs". Thus, he thought "there is no benefit for watching foreign television programs". In contrast, he believed "it is more realistic to watch some Chinese programs [...] concerning what happens around us and in day-to-day life". (Interview BJ_3) Li's perceptions seemed fundamentally based on his own cultural standards and he had not been able to step into a world of foreign culture.

*Case 5: Yuan, a 28-year-old man from Nanjing (bachelor degree)*

He held a deep resistance especially towards romantic movies and television series. He liked to watch their Chinese equivalents, because he "can understand emotion in domestic television programs better". He also explained that there are a lot of differences in the ways of thinking in foreign and Chinese romantic movies. He could "not see everything from the perspective of people in foreign television programs" and understood emotion in both types of romantic movies the same way. (Interview NJ_1)

## 11.1.2 Learners: High Perception with High Acceptance

- **"Previously I did not know them; but now I want to."**

Among all five viewer groups the group of learners is the largest (13 out of the 36 participants). They could perceive some large cultural differences, but they preferred to accept them. For example, Wang said "I do not have any cultural barriers". (Interview SH_3) Likewise, Pan stated "I can accept them". (Interview GZ_5) It seemed that they did not have serious problems accepting cultural differences. They even learned a lot from what they saw in foreign television programs, e.g., holding weddings in church. This could give an impression that these participants as learners were well adapted to watching foreign television programs.

*Case 1: Wang, a 36-year-old advertising agent from Shanghai (master's degree)*

She had watched a great deal of foreign television programs through television and the internet, such as blockbuster movies (e.g., *Ice Age*), television series, documentaries (e.g., *National Geography*), sports program (e.g., *NBA*), and music programs (e.g., *MTV*).

She seemed to have a strong perception of the cultural differences she encountered in foreign television programs. She stated that she could "feel the differences very obvious", such as regards "moral standards", "life", "music", "attitudes towards some sensitive themes", and so on. (Interview SH_3) However, she had more or less accepted various cultural differences at both a surface and a deep level. Generally, she recognized the existence of cultural differences and attempted to understand them, but she did not necessarily agree with them. She said:

> "Some values from foreign countries I cannot agree with, but I try to understand them. I think everything that exists is rational. Since cultural differences exist, I should try to understand them. I must recognize that they really exist; that they are a fact of life. However, I do not have to agree with them. We can both, look for similarities while maintaining dissimilarities between Chinese culture and foreign cultures." (Interview SH_3)

She had an open mind to accepting cultural difference. Her attitude helped her look behind facts to uncover cultural meanings and patterns, and to gain entrance into the cultural worlds of others. In fact, her statements "try to understand them" and "not have to agree with them" might imply that she did not like foreign culture very much; however, her open mind helped her to change her attitudes. Without this ability she would observe other cultures simply as an outsider. Actually, she did profess an acceptance of foreign culture, saying "we can both, look for similarities while maintaining dissimilarities". In addition, she explained that her open mind and receptive attitude were not related to her current residency in an open living environment in Shanghai, because she had grown up in rural China. (Interview SH_3)

She said she did "not have any cultural barriers". Her main purpose in watching foreign television programs was "to broaden my horizons". She seemed to have a slight regret for not having paid much attention to foreign culture during her studies. Therefore, now she would like to watch more foreign programs that help her to broaden her horizons and "to know how other countries look; to know their people, culture, and geography". It was a "learning process" for her "to explore and know more about foreign cultures". Therefore, she summarized her views as follows:

> "No cultural barriers exist. Previously I did not know them, but now I want to. So it is a process of searching for new knowledge." (Interview SH_3)

In addition, the most important criterion for her in selecting and watching foreign television programs was whether she liked them. If she liked them, she would "watch them and not change channels"; if she did not, she would "switch channels immediately". She said: "I wouldn't choose not to watch foreign television programs just because they contained some cultural differences." (Interview SH_3)

*Case 2: Wang, 31-year-old from Beijing (bachelor degree)*

He could perceive some cultural differences: "in general there are no large cultural differences, but there are a few in some areas", e.g., "their habits of communicating with each other", "their language characteristics", "ways of thinking", "ways of life and work", "values", "worldviews". He believed that generally he could mostly accept cultural differences, because he belonged to the young generation born after 1980 that lived in a more open environment. "Many cultural differences in foreign television programs are usual" for him. He explained that during globalization China and foreign countries were gradually "changing" and it was "a process of learning from each other". Many cultural aspects in China were now already "integrated with other foreign culture" e.g., "values and worldviews" and "in general they are the same as us". (Interview BJ_1)

*Case 3: Zhang, a 29-year-old scholar from Xi'An (master's degree)*

He was planning to study in a foreign country. He watched foreign movies, television series, sports programs, and documentaries on television and through the internet. He could definitely feel a lot of "very large" cultural differences. Among them the most important were, e.g., values, beliefs, ways of thinking, and the "superstructure of society". He explained that "perhaps other people do not know these differences, but I know because I study social science". This implied that he had his own special way of looking at cultural difference.

He could not only accept cultural differences, but also liked them very much, because "various cultures are colliding and integrating with each other" thus "cultural differences are becoming smaller". Anyway, he liked foreign culture "relatively more" than Chinese culture, because "their culture is more liberal and our culture is somewhat inflexible and conservative". He liked "everything that is liberal", for example, "their morality" and "wedding ceremonies from Western countries" seemed more open and liberal than in China. (Interview XA_10)

In sum, besides the three cases introduced above, there were still numerous cases between learners. Generally, they could perceive some large cultural differences, but wished to accept them. Pan from Guangzhou said "it is worth learning something from them" and he "looks forward to such a life of freedom" like he had seen in foreign television programs. (Interview GZ_5) Zhou, a 34-year-old woman from Xi'An, could also accept cultural differences at a deep level, because she believed "every culture has its own historical reasons for its formation" and "members of a culture cannot decide how their culture looks". Her statements indicated that every culture had its own rationality. Furthermore, "in the current global environment many Chinese are xenophiles and wish to learn something from foreign countries". However, she criticized the Chinese for just learning "the superficial things from foreign countries" and for not actually understanding "the essence of foreign cultures". (Interview XA_9)

All of these learners have a common character – a cross-cultural mindset based on a simple realization: there are minds out there that think as well as theirs, but differently; all cultures have coherence and an implicit rationale, and no one culture is inherently better or worse than any other. They generally share a set of traits or attitudes as a result of learning, which include:

(1) *Intercultural curiosity*: They have a genuine desire to know more about people and their ways of life, ideals, values and style of dress in other cultures;

(2) *Adaptability*: They have a willingness to try new things in other cultures and to change routines and patterns in their own culture;

(3) *Tolerance and patience*: They are tolerant towards differences and patient at waiting for meanings from other cultures to emerge slowly, as intercultural experiences accumulate.

## 11.1.3 Adherents: Low Perception with Low Acceptance

The attitudes of adherents towards cultural difference are that they cannot perceive any cultural differences that are obvious, but they have a deep resistance to accepting any cultural difference. In this study, there are no adherents (0 participants out of 36), because every interviewee could perceive cultural difference more or less.

## 11.1.4 Adjusters: Low Perception with High Acceptance

- **"No real difference; no problem accepting it."**

The adjusters (10 participants out of 36) were viewers who believed that cultural differences were very small, but they could mostly accept them. They were very interested in differences and glad to see cultural differences in foreign programs, because they wanted to know more about other cultures.

*Case 1: Wu, a 34-year-old manager of an enterprise from Beijing (master's degree)*

He watched all kinds of foreign television programs, e.g., movies, cartoons, and documentaries. He did not perceive any obvious cultural differences: "Cultural differences surely exist, but there is no essential difference. [...] Actually it is difficult to identify real differences". Furthermore, he could accept all kinds of them. He had a sophisticated understanding of cultural difference and believed that he had "no problems accepting them" and nothing was unacceptable to him. (Interview BJ_7)

He explained the social background of globalization for cultural difference in detail. In his opinion "cultural differences are becoming smaller and smaller". Generally the Chinese born before 1980 might still "have some difficulties accepting cultural difference", whereas those born after 1980 "shouldn't have difficulties accepting cultural difference". One reason was that the after-80 generation in China had started to "learn and know a great deal about foreign countries since childhood". They used many things, such as "pencil brands", which come from foreign countries. Furthermore, "the figures in textbooks are all foreigners" and "some articles in Chinese textbooks, such as *The Last Lecture* written by the French author *Alphonse Daudet* come from foreign countries". He liked reading books from Western countries and thought "the ideology in such foreign literatures is very good".

His open attitude toward cultural difference was based on his fundamental belief in global integration, which meant that "the boundaries between countries and culture had gradually become blurred". He said that "Chinese culture is changing according to foreign cultures and foreign cultures are also changing according to Chinese culture to a certain degree", for example, many Chinese are now having "weddings in church" in China; some "business-administration theories from Western countries" which were introduced to China have also changed. He advocated that the Chinese "shouldn't have conservative attitudes towards foreign cultures". Instead, they should, on the one hand, "search for some common ground with foreign countries"; on the other hand, they should "maintain some necessary differences". (Interview BJ_7)

*Case 2: Cai, a 39-year-old media designer from Guangzhou (master's degree)*

He watched all kinds of foreign television programs and was interested in foreign movies the most, such as the genres science fiction, action, romance, and family. He recognized there was "certainly" cultural difference, e.g., "their living environments", "their life habits", and "their attitudes towards life". He said: "now cultural difference is allowed and it is interesting to see some cultural differences". The word "allowed" meant that in his opinion cultural difference was rational. He "enjoys" cultural difference very much. If he "always sees the same faces, it is boring". (Interview GZ_1)

His attitudes towards cultural difference were very open, because as a media designer his work environment was very open and he had had many opportunities to learn about foreign cultures. Furthermore, he had learned something about cultural difference from his teacher in college. He explained:

> "I can often see something from foreign countries. In our media branch, most advertising campaigns and designs are related to foreign cultures or concepts from foreign countries. At school, I attended some lectures about Western and Eastern literature and art comparison. The teachers told us something about cultural difference. Therefore, I can now understand and accept cultural difference very well and enjoy it very much." (Interview GZ_1)

*Case 3: Zhang, a 27-year-old international trader from Guangzhou (master's degree)*

He watched some foreign television programs over the internet, such as American, Korean, and Japanese programs. He particularly liked Korea's television series about love. He could perceive cultural differences in foreign television series, sports programs, and documentaries, e.g., cultural differences regarding characters, narrative style of presenters, drama plots, values, religion, language, and so on. However, he could accept them very easily and even liked them a great deal.

He especially noted the cultural difference of education methods in foreign television. He liked the education methods showed in foreign programs very much, because he believed that foreigner education methods were very open, liberal, and equal. He said "if I have children, I would like to educate my children with the methods I have seen in foreign programs". (Interview GZ_3)

His living environment was very open: "I have a lot of contact with foreigners, not only now, but for the last few years. I have studied in Guangzhou since my childhood. When I was in the upper grades in primary school, I was taught English by foreign English teachers". His education was also not very restricted by his parents, who were not like other typical parents in China: "My parents did not care what I read or watched. I could choose what I want. This book in my hands is a foreign one". (Interview GZ_3)

*Case 4: Yang, a 41-year-old international trader from Guangzhou (bachelor degree)*

She also did not have set attitudes towards cultural difference. She believed that "there is no cultural difference" and she "can accept all of them". Although she did "not have much contact with foreign countries" and did "not know very much about foreign cultures", she did not "want to exclude any cultural differences". She liked to "accept some new things" because "everything that exists is rational and reasonable". (Interview GZ_2)

*Case 5: Shi, a 35-year-old lawyer from Guangzhou (master's degree)*

She had similar opinion to Yang. She watched foreign blockbuster movies and documentaries about nature and daily life in other cultures. She "can accept everything" and "likes their culture very much", because "people in foreign television programs are equal", "their moral standards are very good", and "their education methods are very good". "Good" in her opinion meant "better in accordance with people's instinct, demands, and desire". (Interview GZ_6)

She explained that what she had seen in foreign television programs was not how her teachers and parents had portrayed them, telling her that foreign countries were not good. On the contrary, she thought that "now in our country, on the one hand, we are losing our traditional culture, while, on the other hand, we do not have a modern developed social awareness like in Western developed countries".

She said that her deep acceptance of cultural difference was related to her occupation as a lawyer and her "rebellious and radical" personality. She did not like to "accept orders from other people" and "always accept what other people say", such as what her parents and teachers told her about foreign countries.

*Case 6: Dr. Huang, a 31-year-old college teacher from Shanghai (doctorate)*

He liked watching American classic movies (e.g., *Gone With The Wind, My Fair Lady*), Indian movies, and documentaries (e.g., *Discovery Channel*). He could accept all cultural differences in foreign programs "very naturally", because he "wants to know more about foreign culture". When asked whether he had had much opportunity to learn anything about foreign cultures in his daily life, he replied: "no, not much". He believed that the most important thing to be able to accept cultural difference was to recognize "their rationality":

> "If you watch a Chinese opera from ancient times, the content is, of course, not same as modern life. However, you can still understand the content quite naturally. I do not think the problem is cultural differences. Works on screen, such as movies, television series and operas, if their content is rational and persuasive, there is no problem accepting it." (Interview SH_1)

In sum, the interview results indicated that adjusters with a low perception combined with a high acceptance of cultural difference are relatively rare, even among highly educated and intercultural-experienced participants. This is not because people are willfully prejudiced against cultural difference, but rather because (1) television viewers are not required to make major cultural adaptations when watching foreign television programs. In addition, (2) when have to make major cultural adaptations they often have the ability to "pick and choose", enjoying parts of their intercultural experience by watching foreign programs like and criticizing those they found inconvenient or difficult to deal with.

### 11.1.5 Neutralists: Neutral Perception and Neutral Acceptance

- **"I do not care about cultural difference."**

The neutralists (2 participants out of 36) occupy the middle ground of the four viewer groups. They believed that cultural differences were neither large nor small, and they neither accepted nor did not accept them. They neither paid much attention to cultural difference, nor wanted to adapt it.

Peng, a 25-year-old woman from Beijing, is a neutralist. She said that she did "not care about cultural difference". The most important thing about foreign television programs that concerned her was the "visual effect" which included the camera work, lighting, and editing techniques. (Interview BJ_4) Yu from Beijing held a similar opinion that cultural difference was not important for him. He said, "the most important thing for me is a good drama plot". (Interview BJ_5)

## 11.2 Transition of Television Viewer Types

The transition from escapee to adjuster is a long process based on a continual accumulation of intercultural experiences by watching foreign programming. Figure 14 demonstrates a transmission model of television viewer types. The starting point of the transition is the "escapee" stage. The stages of "adherent", "learner", and "neutralist" are considered the three different middle phases. The end stage is "adjuster".

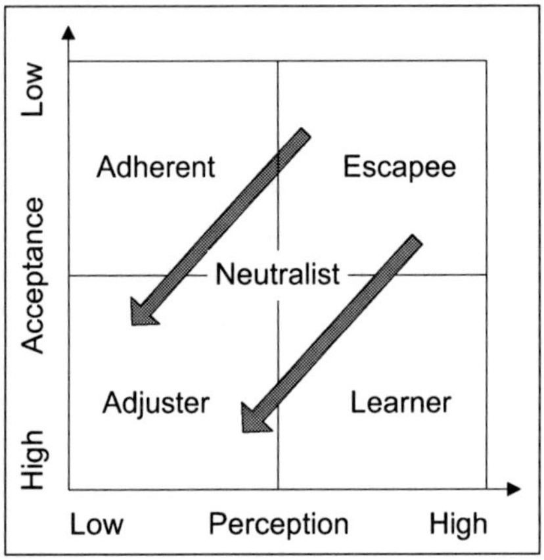


*Figure 14: Transition of Television Viewer Types*

Source: Own illustration.

This gradual transition process was alluded to in some participants' statements:

*Case 1: Wang, 31, from Beijing (bachelor degree)*

He explained his gradual transition process as follows: when he was a child, he did not understand any cultural difference in foreign programming, such as "their way of talking" and "sequence of storytelling". Then, he grew up and "could gradually understand and accept some new things in foreign programs". Now, he could not perceive any large cultural differences. (Interview BJ_1)

*Case 2: Dr. Huang, a 31-year-old college teacher from Shanghai (doctorate)*

He described his understanding of the gradual transition process of cultural difference as follows:

> "Previously, people in China were isolated from the outside world and did not know anything about foreign culture. If they watched something about foreign cultures, they would

be surprised. However, now most people know more or less about foreign cultures and they will not be surprised when they encounter them." (Interview SH_1)

*Case 3: Zhang, a 22-year-old student from Shanghai (bachelor degree)*

In telling a story about his difficulty in getting used to watching foreign music programs (e.g., concerts and operas), he said: "in the beginning I had some barriers to watching them. However, later I became used to listening to foreign music". Now, he watched foreign music programs because he liked the "style of music" in them. (Interview SH_4)

# 12 Summary of Part D2

(1) *Research purpose and questions*: It was proposed that in the minds of television viewers there were many kinds of cultural differences. Some stand out, because they somehow seemed important or obvious to viewers.

Thus, this personal-interview part of the research with Chinese viewers primarily aims to explore cultural-difference dimensions in foreign television programs, which will be further quantitatively evaluated in a survey of Chinese viewers, because only interpretations from Chinese viewers can be understood well by Chinese viewers. This part has two research questions:

- What kinds of cultural difference can Chinese TV viewers perceive? This question aims to construct a catalogue in order to summarize various cultural-difference dimensions systematically based on the statements and interpretations of participants.

- To what extent can they perceive and accept such cultural difference? This aims to categorize viewer types according to degree of perception and acceptance.

(2) *Method*: A total of 36 interviews with Chinese television viewers with different demographic characteristics were conducted in October and November 2009 in five sample cities in China. However, due to the low number of participants the representatives of the sample as well as the comprehensiveness of cultural-difference types cannot be guaranteed.

(3) *Results*: The first part of results concerns about various *cultural-difference dimensions*. The participants' statements underlined the facts that:

- *Clear boundary between "their" and "our" culture*: Participants classified the foreign culture on the side of "openness and liberty", and the Chinese culture on the side of "conservatism".

- *Cultural-difference dimensions*: Their common understanding about cultural difference consists of many cultural dimensions. Based on numerous statements of participants a total of 10 sets of dimensions/42 dimensions were found and summarized in a catalogue. They were divided into two groups: genre-*related* cultural differences (incl. movie & TV series, sport program, documentary, entertainment show, news program, and music program) and genre-*unrelated* cultural differences (e.g., language, value, ways of life, and education method).

- *Television viewer types*: According to degree of (high vs. low) perception and (high vs. low) acceptance of cultural difference, all participants were divided into five viewers groups: escapees, learners, adherents, adjusters, and neutralists. The transition from escapee to adjuster is a long process based on continual accumulation of intercultural experience. The start point is “escapees”. “Adherents”, “learners”, and “neutralists” are middle phases. The end stage is “adjusters”.

# Part D3: Survey – Perception and Acceptance of Cultural Differences

# 13 Perception and Acceptance of Cultural Difference

## 13.1 Method: Survey with Chinese Television Viewers

### 13.1.1 Design of Survey

The survey consists of three parts - introduction, key part, and personal information: (See Appendix 4: Survey)

(1) *Introduction*: At beginning of survey some definitions about foreign television programs are made: "The foreign television programs" include foreign movies and television series, sports programs, documentaries, entertainment shows, news programs, and music programs, which are broadcast by Chinese television stations or supplied by internet Websites. They are produced by foreign media corporations, such as the United States, European countries (e.g., Britain, Germany, and France), Asian countries (e.g., Japan and Korea). The Chinese indigenous television programs or programs from Hongkong, Taiwan, and Macau do not belong to foreign TV programs.

(2) *Key part*: This part aims to evaluate the perception and acceptance degree of 10 sets/42 single cultural-difference dimensions which derived from the results of television schedule analysis and personal interviews. The first six sets are genre-related cultural differences in the following six television genres – movies & television series, sports programs, documentaries, entertainment shows, news programs, and music programs. The last four sets are genre-unrelated cultural differences – language, values, ways of life, and education methods. Each set of dimensions was measured using between two and seven separate dimensions.

Each set of genre-*related* dimensions has four questions about:

- Viewing foreign television programs;
- Perception and acceptance of cultural difference in this genre;
- Preference of country of origin;
- Preferred and not-preferred foreign programs in this genre.

Each set of genre-*unrelated* dimensions has only one question about:

- Perception and acceptance of cultural difference.

(3) *Personal information*: This includes participants' gender, age, education, income, city of residence, and their knowledge about foreign culture. These measures are used as influential factors to examine their impact on their perception and acceptance of cultural difference.

## 13.1.2 Measures and Analysis Model

### 13.1.2.1 Perception and Acceptance Index

Various cultural-difference dimensions are measured using the *perception index (PI)* and *acceptance index (AI)*. As discussed above, perception and acceptance are two different concepts. They vary between cultural dimensions and television viewers. Using only one of them would not be sufficient for adjudging viewers' opinions about cultural difference. Both indices carry out a quantitative evaluation of each cultural dimension and make a comparison between them possible. The size of the indices can determine how "big" and "acceptable" each cultural dimension is.

To evaluate the perception and acceptance of cultural difference the participants are asked the following: "To what extent could you perceive the following cultural differences? To what extent could you accept them?" Perception index (1 = very small to 5 = very large) and acceptance index (1 = totally acceptable to 5 = not acceptable) were scored on a 5-point scale, which clearly defines the cut-off point of large and small cultural difference. Both quantitative indices indicate the difference/similarity between culture in foreign television programs and Chinese culture. However, PI indicates the size of the difference/similarity, while AI indicates their acceptance.

### 13.1.2.2 Cultural-Difference Dimension

All of the cultural-difference dimensions in the survey were derived from the analysis results of the personal interviews. In the survey each dimension was interpreted in a relatively implicit rather than an explicit way. Perhaps some scholars might think that they are not precise enough for interpreting the questions and each participant may have a different understanding of them.

There are two reasons for the implicit interpretations of each dimension in the survey: (1) Chinese culture is a high-context culture. People do not need much explicit and detailed information to understand each other. Interpretations, which are too complicated, and redundant information that they do not need will not only confuse them, but also make them impatient. (2) Some participants just have a primary-school or middle-school level of education. The requirement to read complicated information is too high for them.

In order to examine whether the interpretations in the survey are too complicated or contain redundant information for Chinese participants, 30 pre-test surveys were conducted on Chinese participants. Based on their opinions, the author modified the interpretations about cultural difference in the survey, which include only the information they needed. In the final version of the survey each dimension was described using a few concise keywords.

### 13.1.2.3 Influential Factors

The influential factors in this study include demographics (gender, age, education, and income), knowledge about foreign culture, viewing foreign television programs, cultural proximity, and city of residence. All were included in the analyses as control factors.

- **Demographic (gender, age, education, income)**

*Gender* aims to examine the difference of opinions between male and female participants. Gender was coded with male equal to 1 and female equal to 2.

*Age* aims to examine the difference between different generations. There are three age groups in this survey: 20-29, 30-39, and 40-49. It is particularly important to examine differences between the age groups 20-29 and 30-39. As introduced above, *China's Opening Up Policy* started in 1978. The age group 20-29 is the generation born after 1980 and is the so-called *post-1980s generation.* The age group 30-39 is the generation born before 1980. Both generations have large generation gaps, because their living environments are quite different. People younger than 20 and older than 50 are not the target group of this study, because they are very difficult to recruit and this survey is also too difficult for them to fill out.

*Education* is evaluated by asking participants to report their reached highest education level. It has four groups: primary/middle school, bachelor, master, doctorate.

*Income* is evaluated by respondents' monthly average household income by selecting from four categories ranging from ¥0-999, ¥1000-4999, ¥5000-9999, and over ¥10000. In China, education level and income are not proportional because a high education level does not indicate a high income. For example, although many scholars in China have the title doctorate, they may have a much lower salary than people who work in enterprises who only reached a middle-school level of education.

- **Cultural exposure (knowledge, viewing)**

*Knowledge*: Participants were asked "How much do you know about foreign cultures?" They used a 5-point scale (1 = very much to 5 = very little) to answer this question.

*Viewing*: Equally, the measurement about viewing foreign television programs has a same 5-point scale (1 = very much to 5 = very little). The question "how much foreign movies and television series/sports programs/documentaries/entertainment shows/news programs/music programs do you watch?" is always the first in every question block for each television genre, because it could help them remember some of the foreign programs they have seen and prepare them for the following questions about cultural difference. This question about the amount of foreign programs they watch is repeated for each television genre.

- **Cultural Proximity**

To examine cultural proximity the participants were asked "how much do you like this television genre from these countries (except China)?" This question is also repeated for each genre. All of the foreign countries are divided into four groups: North America (e.g., U.S.), Europe (e.g., Britain), Asia (e.g., Japan and Korea), and other foreign countries. Respondents were asked to indicate on a 5-point scale (1 = very like to 5 = not like) how much they like a certain genre from culturally proximate and culturally distant country groups.

- **City of residence**

Five resident cities are also used as an influential factor to examine possible regional differences. They were coded with Beijing equal to 1, Shanghai 2, Guangzhou 3, Nanjing 4, and Xi'An 5. The five numbers also indicate the integrated ranking of five cities according to the following four criteria: administrative level, degree of economic development, openness to outside world, and quantitative supply of foreign television programs. "1" is the highest and "5" is the lowest ranking. (See Table 15)

*Table 15: Ranking of Fives Sample Cities*

| Code | City | Admi. level | Economy | Openness | TV Supply |
|---|---|---|---|---|---|
| 1 | Beijing | Center | High | High | 354.2 hours |
| 2 | Shanghai | Province | High | High | 216.1 hours |
| 3 | Guangzhou | Province | High | High | 147.0 hours |

| | | | | | |
|---|---|---|---|---|---|
| 4 | Nanjing | Province | Middle | Middle | 129.1 hours |
| 5 | Xi'An | Province | Middle | Middle | 50.8 hours |

Note: Admin. level = Administrative level. Quantitative supply of foreign television programs based on the results of television schedule analysis.
Source: Own illustration.

- **Program preference (open question)**

In each question block of six genres there are two open questions: "Which foreign TV programs do you like to watch?" and "Which foreign TV programs do you not like to watch?" These two questions were repeated for each genre, because the answers to both might be quite different depending on television genre. Both questions have two aims:

- To examine whether participants understand the questions in the survey correctly.
- To describe a framework of foreign TV programs in the minds of viewers.

#### 13.1.2.4 Analysis Model

Figure 15 demonstrates the analysis model in this part of the research, which includes two main indices (perception index and acceptance index) and five influential factors which include demographics (gender, age, education, and income), knowledge about foreign culture, viewing foreign TV programs, cultural proximity, and city of residence.

### 13.1.3 Sample Description

For the survey, the television viewers with different demographic characteristics in five sample cities (Beijing, Shanghai, Guangzhou, Nanjing and Xi'An) were the target population. Surveys were conducted on Fridays, Saturdays, and Sundays in May and June 2010 in China. (See Table 16) A convenient sample of 450 Chinese viewers was recruited. The selection of different sample cities was made in order to examine possible regional difference.

The participants were selected socio-demographically in order to capture a wide variety of viewer types, who might hold quite different opinions about cultural difference. The total sample size of the final survey was 450. Fifty-three percent of participants were females and 45% were males. Their ages ranged from 20 to 49, with 83.3% between the ages of 20 and 29, with 12.4% between the ages of 30 to 39. It is very difficult to reach people over 40 and the percentage of the sample in that age group, 40-49, was limited to 2.4%. The education level ranged from primary school to doctorate, with 21.8% primary/middle school, 70.1% bachelor degree and 8.1% (incl. doctorate) master or over. Income ranged from ¥ 0-999 to over ¥ 5000, with 23.3% under ¥ 999, and 58.7% between ¥ 1000 and 4999. Each sample city had 84 to 99 samples. (See Table 17)

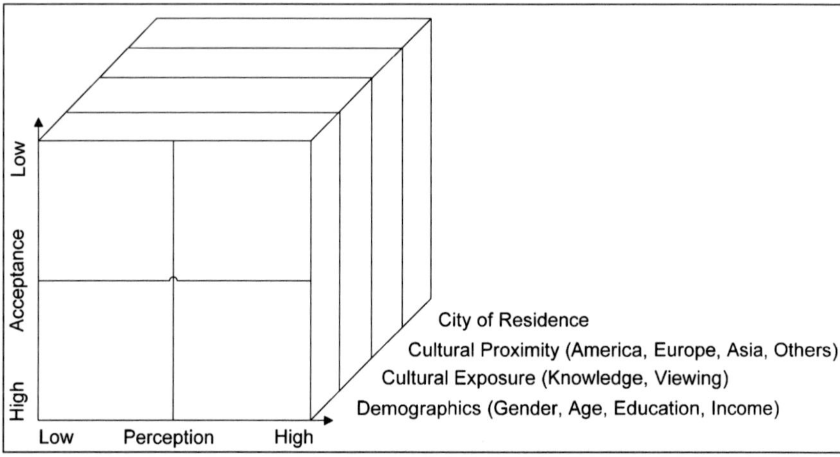


*Figure 15: Analysis Model*

Source: Own illustration.

*Table 16: Date of Survey Conduction*

| **Sample City** | **Date** | **Day** |
|---|---|---|
| Beijing | 14 – 16 May 2010 | Friday, Saturday, Sunday |
| Xi'An | 21 – 23 May 2010 | Friday, Saturday, Sunday |
| Nanjing | 28 – 30 May 2010 | Friday, Saturday, Sunday |
| Shanghai | 04 – 06 June 2010 | Friday, Saturday, Sunday |
| Guangzhou | 18 – 20 June 2010 | Friday, Saturday, Sunday |

Source: Own illustration.

Table 18 shows the statistics of population allocation in 2010 in China. These data were obtained from the *National Bureau of Statistics of China*. They are the unique official population data in China; however, they are not detailed enough for comparison with the sample allocation in this study. For example, there are just three age groups: 0-14, 15-59, and over 60. People do not know exactly how the age group 15-59 is further allocated. There is no information about income. Therefore, the representativeness of this study cannot be judged. Furthermore, population allocation varies significantly across regions and cities in China.

*Table 17: Sample Allocation in Five Sample Cities*

| | | City | | | | | Total | | | |
|---|---|---|---|---|---|---|---|---|---|---|
| | | BJ | SH | GZ | NJ | XA | N | % | Val. | Mis. |
| Gender | Male | 37 | 44 | 41 | 36 | 44 | 202 | 44.9 | 441 | 9 |
| | Female | 59 | 34 | 45 | 48 | 53 | 239 | 53.1 | | |
| Age | 20-29 | 66 | 61 | 76 | 81 | 91 | 375 | 83.3 | 442 | 8 |
| | 30-39 | 26 | 14 | 8 | 3 | 5 | 56 | 12.4 | | |
| | 40-49 | 4 | 3 | 3 | 0 | 1 | 11 | 2.4 | | |
| Education | Pri./mid. school | 6 | 20 | 46 | 9 | 15 | 96 | 21.8 | 441 | 9 |
| | Bachelor | 76 | 50 | 39 | 68 | 76 | 309 | 70.1 | | |
| | Over master | 14 | 8 | 1 | 7 | 6 | 36 | 8.1 | | |
| Income | ¥ 0-999 | 9 | 14 | 18 | 24 | 40 | 105 | 23.3 | 424 | 26 |
| | ¥ 1000-4999 | 63 | 48 | 59 | 51 | 43 | 264 | 58.7 | | |
| | Over ¥ 5000 | 24 | 14 | 6 | 5 | 6 | 55 | 12.2 | | |
| City | Total | 96 | 84 | 87 | 84 | 99 | 450 | 100 | 450 | 0 |

*Note: BJ = Beijing, SH= Shanghai, GZ = Guangzhou, NJ = Nanjing, XA = Xi'An.*
*Source: Own illustration.*

*Table 18: Population Allocation in China in 2010*

| | | Population in China (%) | Sample in this Study (%) |
|---|---|---|---|
| Gender | Male | 51.3 | 44.9 |
| | Female | 48.7 | 53.1 |
| Age | 0-14 | 16.6 | 0 |
| | 15-59 | 70.1 | 98.1 |
| | Over 60 | 13.3 | 0 |
| Education | Middle school | 51.6 | 21.8 |

| | | | |
|---|---|---|---|
| | Primary school | 26.2 | |
| | College | 8.7 | 78.2 |

Source: NBSC, 2010.

All surveys were conducted in five central public parks/squares located in the center of each sample city. These five central public parks and squares are:

- Beijing: *Di Tan Park* (Temple of Earth Park) – a large public park in the northeast part of the central area in Beijing. It was previously a temple for worshipping the emperors;
- Shanghai: *Ren Min Square* (Peoples' Square) – a large public square in the middle of the central area in Shanghai and nearby Nanjing Road (the largest shopping district in Shanghai);
- Guangzhou: *Yue Xiu Park* – a large public park in central Guangzhou;
- Nanjing: Xin Jie Kou Square – a large public square in central Nanjing (the largest shopping district in the city);
- Xi'An: *Xing Qing Gong Park* – a large public park in the southeast part of the central area of Xi'An. It is one of the oldest parks in Xi'An and is considered an historical and cultural site.

All of the sites are located in the city center or nearby. Various people from various districts go to public parks or squares for different reasons. Some people do sports activities there, such as Tai Ji, or dancing, while others go to fly kites or play chess. It is also a place for meeting with friends to chat or somewhere to rest after shopping; it is also a popular area for couples to meet on a date. Some people work nearby and thus have breakfast, dinner, or just a drink there. Families go there to take a walk or to play, e.g., those with young children. In public parks and on public squares various people of different ages, educational backgrounds, and occupations can be found.

People there to take a rest are generally not in a hurry, and there are some comfortable chairs and benches allowing people to sit in the sunshine. Therefore, the survey participants were able to sit down and fill it out in these locations. The areas used for the survey were also very quiet as they were set away from the noise of the streets. This reduced the risk of participants being distracted so they could concentrate on the survey. The author waited with each participant until they had finished the survey and to answer any questions. Thus, the return rate was over 90%.

### 13.1.4 Summary of Foreign Television Genres in China

Before the results of survey are given, the various genres of foreign television programs should be summarized first. Only through providing this overview of television genres can people know exactly what kinds of foreign programs the research results are concerning. All of the genres introduced in following are based on the results of the three research parts:

(1) *Pre-study television schedule analysis*: The types of foreign programs broadcast in schedule from 37 television channels were summarized.

(2) *Personal interview*: The types of foreign programs, which were interpreted by 36 Chinese television viewers, were summarized.

(3) *Open questions in the survey*: The types of foreign programs were summarized based on the answers given by 450 Chinese viewers to the two following open questions: "Which foreign television programs do you like to watch?" and "Which foreign television programs do you not like to watch?"

The results of the two open questions in the survey indicated that the types of foreign television programs found in the television schedule were just a small part of the whole offering of foreign programs in China. Many television viewers watched foreign programs not only on television, but also through the internet. The results of the two open questions also indicated that most participants knew various foreign television programs and had correctly understood the questions in the survey.

On the whole, the results of all three research parts about foreign television genres indicated that the most-well-known foreign programs in China were primarily universal television programs. They were produced by foreign media corporations for the global market, not just for particular national markets.

- **Movies**

(1) *America*: The provision and consumption of foreign movies was concentrated on Hollywood movies from the United States. They had a broad range of themes; the most popular being science fiction (e.g., *Star Wars*), action (e.g., *James Bond*), and children's animation (e.g., *Kings of Lion*). Further types such as disaster movies (e.g., *Titanic*), fantasy (e.g., *Harry Potter*), war, romance, horror (e.g., *The Silence of the Lambs*), vampires, adventure, music, and crime were also supplied and consumed in large numbers. Some classic Hollywood movies (e.g., *Gone with the Wind* and *Roman Holiday*) were still remembered by many Chinese TV viewers. (See Table 19)

(2) *Europe*: The movies from European countries were obviously less well known compared to U.S. movies. They were mostly chick-flick movies, for example, the Italian movies *Cinema Paradiso* and *Life is Beautiful*, and the British

comedy *Mr. Bean*. Movies from other European countries, such as France, Germany and Spain, were rarely mentioned by Chinese viewers.

(3) *Asia*: The movies from Asian countries were primarily Japanese animated movies and Korean horrors. The most famous Japanese movies in China were mostly animated, such as *Castle in the Sky* and *Spirited Away*, which were produced by Japan's famous master of animation *Hayao Miyazaki*. The Korean movies mentioned most often were horror movies, e.g., *Bunshinsaba* (Ask The Spirits) and *Bunhongshin* (The Red Shoes). The most popular Indian movies were primarily about weddings, such as *Namastey London*.

*Table 19: Types of Foreign Movies*

| | |
|---|---|
| (1) American | |
| Science fiction | *Spider-Man, The Matrix, Star Wars, X-Men, Superman, E.T., Alien* |
| Action | *James Bond, Zorro, True Lies, Charlie's Angels* |
| Animation | *Shrek, Kings of Lion, Avatar, Transformers, Ice Age, Kung-Fu Panda, Madagascar, Atlantis, Cinderella, The Little Mermaid* |
| Disaster | *Titanic, 2012* |
| Fantasy | *Harry Potter, Lord of The Rings, Pirates of the Caribbean* |
| War | *The Patriot, Pearl Harbor, Tora! Tora! Tora!, Brave Heart* |
| Horror | *The Silence of the Lambs, Scary Movie* |
| Vampire | *Blade, Twilight, Daybreakers* |
| Romance | *Mr. & Mrs. Smith* |
| Adventure | *Indiana Jones, Tomb Raider, The Mummy* |
| Musical | *High School Musical* |
| Children | *102 Dalmatians, Home Alone, Charlie and the Chocolate Factory, Alice in Wonderland* |
| Classic | *Gone with the Wind, Waterloo Bridge, Roman Holiday* |
| Modern classic | *The Godfather, Sleepless in Seattle, The Forrest Gump* |
| (2) European | |
| Chick flick | Italy: *Cinema Paradiso, Life is Beautiful, The Legend of 1900* |
| Humor | Britain: *Mr. Bean* |

| (3) Asian | |
|---|---|
| Cartoon | Japan: *Castle in the Sky*, *My Neighbor Totoro*, *Spirited Away* |
| Horror | Korea: *Bunshinsaba*, *Dark Forest – Horror Tales*, *Bunhongshin* |
| Wedding | India: *Namastey London* |

Source: Own illustration.

- **Other television genres**

*(1) Television series*

*America*: Although the television series from the United States were very popular in China, they did not dominate because a lot of Asian television series also were supplied and consumed at an equal level to the American ones. The popular U.S. television series were, e.g., the medical drama *Grey's Anatomy*, the female drama *Desperate Housewives*, the vampire series *The Vampire Diaries*, the comedy *Friends,* the cartoons *Tom and Jerry* and *Sponge Bob*, as well as the children's show *The Teletubbies*. (See Table 20)

*Asia*: The Asian television series in China were primarily cartoons and romantic series. A large proportion of the cartoons had a sporting theme, for example, racing, basketball, baseball, football, and tennis. The most famous Japanese cartoon series in China were e.g., *Detective Conan*, *Sakura Momoko*, and *Doraemon*. Two classic Japanese romantic series were *Love Letter* (QingShu) and *Tokyo Love Story* (DongJing AiQing GuShi). The most popular Korean series were usually also concerning romance, e.g., *Full House* (LangMan ManWu). Some others were comedies, such as *Funny Family* (GaoXiao YiJiaRen).

*(2) Sports programs*

Most foreign sports programs concerned commercial sports, such as the *NBA* (National Basketball Association), the *UEFA Champions League* (United Europe Football Association), the *WWE* (World Wrestling Entertainment), boxing, golf, tennis, formula 1, and snooker. The other non-commercial sports programs were not popular in China, such as swimming, gymnastics, badminton, and volleyball. (See Table 20)

Besides these sports types with a global appeal, some countries had their own special sports programs with a long history of sports culture, for instance Spanish bull fighting; European equestrian and polo; Japanese baseball, sumo, and judo; and Indian cricket.

Well-known foreign sports channels in China were *NCAA* (National Collegiate Athletic Association) and *ESPN* (Entertainment & Sport Program Network). A popular sports magazine program was *Marlboro Sports World.*

*(3) Documentaries*

The most well-known foreign documentary channels in China are *Discovery*, *National Geographic*, the *BBC* (British Broadcasting Corporation), and *NHK* (Nippon Hōsō Kyōkai) documentaries. The themes of the documentaries were varied, including geography, animals, the environment, war, adventure, science, humans, history, and so on. (See Table 20)

*(4) Entertainment shows*

*America*: The popular U.S. entertainment shows were, for example, the talent shows *American Idol* and *American's Next Top Model*, the reality shows *Survivor* and *Man vs. Wild*, the game shows *The Moment of Truth* and *Do not Forget the Lyrics*, the talk show *Oprah*, and the science show *Myth Busters* by Discovery. (See Table 20)

*Asia*: The Japanese entertainment shows were primarily the creativity show *Super Change* (ChaoJi BianBianBian), the variety show *Smap X Smap*, and the talent show *The Red White Song Competition* (HongBai GeSai). The Korean entertainment programs were mostly variety shows like *X-man*, *Love Letter* (QingShu).

*Europe*: The entertainment shows from Europe were, for example, the British reality show *Big Brother*, the talent show *Britain's Got Talent*, and the dating game show *Take Me Out*, as well as the French sports game show *Games without Frontiers*.

*(5) News programs*

News programs as a television genre are different to other foreign television genres, because foreign news programs are subject to the most censored and are prohibited from being broadcast as an independent program on Chinese television channels. Viewers can only watch some of them via the internet or satellite television in hotels. However, a lot of the participants were very familiar with foreign news channels, such as *CNN* (The Cable News Network) from the United States, *BBC* (British Broadcasting Corporation) from Britain, *Les Guignols de L'info* from France, *KBC* (Korean Broadcasting Commission) from Korea, and *Asahi Shimbun* from Japan. (See Table 20)

*(6) Music programs*

Foreign music programs included hit-music shows, concerts, music events, music-discussion programs, operas, dance music, and singing competitions. The hit-music shows from the United States, Japan, and Korea were the most popular in China compared to those from other countries, such as *Music Bank* from Korea. The most well-known foreign concerts were classical concerts, such as *The New Year's Concert of the Vienna Philharmonic*, the *Beethoven Symphony Concert*, and pop-music concerts from Japan and Korea. The music events were, for example, the *Grammy Awards*, and the *Dublin Music Festival*. The Irish traditional dance music of *River Dance* was also very well liked by Chinese viewers. There were still some cultural special music, such as U.S. rock and country music, Korean street music, and African drum music. (See Table 20)

*Table 20: Types of Other Television Genres*

| (1) TV series | |
|---|---|
| America | Medical: *Grey's Anatomy* |
| | Women: *Desperate Housewives, Gossip Girl* |
| | Suspense: *Lost* |
| | Prison: *Prison Break* |
| | Vampire: *The Vampire Diaries* |
| | Horror: *Saw* |
| | Comedy: *Big Bang Theory, Friends, Growing Pains* |
| | Cartoon: *Tom and Jerry, Mickey Mouse, Sponge Bob* |
| | Children: *Teletubbies* |
| Japan | Cartoon: *Detective Conan, Sakura Momoko, Doraemon* |
| | Romance: *Love Letter, Tokyo Love Story* |
| Korea | Romance: *Full House, The Princess Diaries* |
| | Comedy: *Funny Family* |
| (2) Sports programs | |
| Commercial | Commercial: *NBA, UEFA Champion League, WWE*, boxing, golf, tennis, formula 1, snooker |
| | Non-commercial: swimming, gymnastics, badminton, volleyball |
| Special sports | Spain: bull fighting |
| | Europe: equestrian, polo |
| | Japan: baseball, sumo, judo |
| | India: cricket |
| Channel | *NCAA, ESPN* |
| Magazine | *Marlboro Sports World* |

| (3) Documentaries | |
|---|---|
| Channel | *Discovery*, *National Geographic*, *BBC*, *NHK* |
| Theme | Geography, animal, war, adventure, science, culture, history |
| **(4) Entertainment shows** | |
| America | Talent shows: *American Idol*, *American's Next Top Model* |
| | Reality shows: *Survivor*, *Man vs. Wild*, *The Apprentice* |
| | Game shows: *The Moment of Truth*, *Do not Forget the Lyrics* |
| | Talk shows: *Oprah* |
| | Science shows: *Myth Busters* (by *The Discovery Channel*) |
| Asia | Japan: the creativity show *Super Change*, the variety show *Smap X Smap*, the talent show *The Red White Song Competition* |
| | Korea: the variety shows *X-man*, *Love Letter* (QingShu) |
| Europe | Britain: the reality show *Big Brother*, the talent show *Britain's Got Talent*, the dating game show *Take me out* |
| | France: the game show *Games without Frontiers* |
| **(5) News programs** | |
| America | *CNN* |
| Asia | Japan: *Asahi Shimbun* |
| | Korean: *KBC* |
| Europe | Britain: *BBC* |
| | France: *Les Guignols de L'info* |
| **(6) Music programs** | |
| Format | Hit-music shows: *Music Bank* |
| | Concert: *The New Year's Concert of the Vienna Philharmonic* |
| | Music events: *the Grammy Awards*, *the Dublin Music Festival* |
| | Dance music: *River Dance* |
| Special music | U.S.: rock and country music |
| | Korea: street music |
| | Africa: drum music |

Source: Own illustration.

## 13.2 Research Purpose, Research Questions, and Hypothesis

The previous part of the research, the personal interview, has answered the question regarding what types of cultural difference could be perceived by Chinese television viewers. On their own the results of the personal interviews would not reveal what central each type of cultural difference is like for Chinese viewers when watching foreign programs, because the interview approach is not able to measure and control all relevant variables statistically, which might make various difference dimensions and viewers even roughly comparable.

In this part it is proposed that cultural difference across the entire range of possible dimensions should not be perceived and accepted uniformly to the same degree. For example, acceptance might thus appear higher in regard to a superficial difference (e.g., appearance of movie characters) than a difference at a deep level (e.g., worldview).

To examine the perception and acceptance of cultural difference reliably and empirically, this study developed a quantitative method through a survey to measure the degree of the perceived and accepted cultural differences. As for the personal interview, the surveys were also evaluated by Chinese television viewers, because watching foreign television programs finally occurs at an individual level. All cultural-difference dimensions in the survey, which were evaluated by Chinese participants, were formulated based on their statements in the interviews. "Cultural difference" in the survey means the differences between cultures in foreign television programs as "their" culture and Chinese culture as "our" culture for Chinese viewers.

Three central research questions are developed to examine cultural differences:

(1) To what extent could Chinese viewers perceive and accept various cultural differences in foreign television programs?

(2) Which stages of cultural adaptation are Chinese viewers currently in?

(3) How large is the impact of influential factors on the perception and acceptance of cultural difference?

### 13.2.1 Perception and Acceptance of Cultural Differences

- **Cultural convergence vs. cultural difference**

*Cultural convergence*: To achieve advantages of scale many media companies produce universal television programs to target the shared habits and tastes of particular market segments at global level, rather than national audiences on the basis of geographical proximity. A new universal television culture aspires to a borderless world in which people transcend different cultures and understand each other mutually. (Morley & Robins, 1997, p. 12) Thus, there is a belief in the world cultural convergence of lifestyle, culture and behavior among audience segments across the world. Some countries are worried that cultural, social, and political values and ideologies in the media products from Western developed countries will supposedly drive out the more fragile native or traditional cultural values in developing countries. (Berger, 1992, p. 14) This is known as "cultural imperialism".

*Cultural difference*: The global flow of universal media products by no means indicates universal reception and the end of difference. (Cohen, 1992, p. 32) Despite similar appearances on the surface, real differences remain and will probably remain that way for quite some time. Television programs rooted in one culture still have a more or less diminished appeal elsewhere, as viewers find it difficult to identify some unfamiliar or unexperienced cultures. (Hoskins et al., 1997, p. 32) Although media corporations make an effort to appeal to a large number of audience through universal television, actually not everyone feels attracted to this kind of universal culture. Most can still feel cultural barriers between their own and the universal television culture.

Thus, the theoretical discussions in this issue raise the following question: Which tendency dominates in the minds of television viewers – cultural difference or cultural convergence? The essence of this question concerns the degree of perceived and accepted cultural differences in foreign television programs.

In the survey, the concepts "perception" and "acceptance" of cultural difference are used to measure the degree of cultural convergence. They indicate two different aspects of viewers' attitudes towards cultural difference: *Perception* indicates the size of cultural difference, in other words, how large cultural difference is perceived by viewers. This may be small or large. *Acceptance* indicates to what extent viewers see other cultures as valid. This varies between acceptable and not acceptable.

Thus, based on reasoning drawn from theories about culture difference in media products, the following hypotheses are proposed:

H1a: TV viewers perceive the cultural differences to be large.

H1b: TV viewers accept the cultural differences mostly.

- **Genre-related cultural difference**

The comparison of cultural difference between television genres should be made because what might be the case for fiction (e.g., movies) may not necessarily be so for non-fiction (e.g., news and documentaries).

The cultural barrier is likely to be lower for works with little linguistic content than for works in which speech is an important element. (Morley & Robins, 1997, p. 62) For this reason, programming with a high speech content is unpopular with foreign viewers because it is subject to a great deal of misunderstanding. Thus, informational programming (e.g., news) will encounter language barriers as the appeal of viewing is reduced by the need to listen to dubbed voices or read subtitles, and further by the difficulty in understanding unfamiliar accents. Conversely, a channel such as *MTV* builds its international appeal precisely on the fact that it has almost no problems regarding language, and words are practically irrelevant. Programming that is not dependent on understanding language will attract a widely dispersed and sometimes potentially large audience. Therefore:

> H2a: The genres with little linguistic content (music programs) have smaller cultural difference than those in which speech is an important element (news).
> H2b: The genres with little linguistic content (music programs) are more acceptable than those in which speech is an important element (news).

- **Genre-unrelated cultural difference**

*(1) Language*

Language is a "natural" barrier to accepting foreign television programs. Language is a complex and troublesome vehicle affecting both, cultures and viewers involved in meaningful exchange (Schmidt, 2007, p. 101), since it provides the conceptual categories to influence how its speakers' perceptions are encoded and stored (Jandt, 1998, pp. 169-170). Language difference might be a major factor in determining cultural barrier by watching foreign television programs, because viewers' feelings of cultural affinity for programs depend on language more than any other cultural element. This linguistic obstacle has to be negotiated through dubbing or subtitling when television programs are introduced to audiences speaking other languages. However, these interventions might disrupt their viewing pleasure.

Furthermore, the results of television schedule analysis indicated that the original language in almost all foreign television programs must be modified significantly. 51.8% were dubbed and the remaining 47.8% must be replaced by a new spoken dialogue in the destination language, which was written by indigenous editors or presenters. Only .4% programs were broadcast in their original language with or without subtitles. Therefore, it is proposed that:

H3a: Language difference is the difference that is the most difficult to accept.
H3b: Viewers prefer watching foreign television programs with dubbing rather than those in the original language with subtitles.

*(2) Education method/way of life*

The results of the personal interviews indicated that cultural differences of "education method" and "way of life" were perceived as being large, but they were easy to accept. Some participants even preferred the education method and way of life that was presented in foreign television programs over Chinese ones. Thus, both dimensions have a high perception degree and a high acceptance degree.

*Education method*: Chinese participants thought that foreign education methods were more open and freer than Chinese ones. Thus, people in foreign television programs seemed independent. In contrast, the Chinese were seen as dependent on each other and must show obedience to parents, elders, or superiors. Some participants would like to educate their children using the methods they had seen in foreign programs.

*Way of life*: This includes numerous daily routines, customs, and rituals. Many participants liked and accepted the Western way of life even better than the Chinese one. For example, they would like to have their wedding in a church just like they have seen in foreign programs. They would like to have dinner or meet friends in Western cafes, restaurants, and pizzerias. The way of life from Western developed countries is now a new mode of lifestyle in China. Therefore:

H4a: The distance between the perception and acceptance indices of education method is larger than other cultural dimensions.
H4b: The distance between the perception and acceptance indices of way of life is larger than other cultural dimensions.

- **Cultural level**

Another perspective for making a comparison between particular dimensions possible is cultural level. In the "catalogue of cultural-difference dimensions" depicted in Table 13 all dimensions were divided into three levels: surface, middle and deep. However, the surface level can be modified over time and is easier to accept, such in regard to the dress style of characters in movies. However, uncovering and deciphering a culture dimension at a deep level, such as worldview, is difficult because they are not accessible at the surface attitudinal level and understanding it requires a long learning process. Thus, the following is proposed:

> H5a: The cultural difference at a deep level is perceived to be larger than the cultural difference at superficial level.
>
> H5b: The cultural difference at a deep level is less acceptable than the cultural difference at a superficial level.

## 13.2.2 Stages of Cultural Adaptation – Television Viewer Types

The measure and comparison of particular cultural-difference dimensions did not indicate the stages of the intercultural adaptation process of television viewers. In the results of personal interviews, all participants were divided into five viewer types according to their perception and acceptance of cultural difference: escapees, learners, adherents, adjusters, and neutralists. (See Figure 14)

Since the launch of China's *Reform and Opening-Up Policy* in 1978, Chinese society has gradually been opened up to the outside world. On the one hand, the young generations in China after the 1980s are learning more and more from Western developed countries. On the other hand, traditional Chinese culture is gradually being lost in the minds of young people in China.

The results of the personal interviews indicated that most participants who watched a lot of foreign programs were young. They could still perceive some large cultural differences. However, they kept an open mind towards learning cultural difference by actively watching foreign television. They neither wanted to escape from cultural difference, nor adhere to Chinese culture. They would like to be part of globalization. Thus, they felt cultural difference was getting smaller and smaller. In the interviews, the participants who kept an open mind towards cultural difference were usually well educated and had a great deal of knowledge about foreign culture and intercultural experiences. Therefore:

> H6a: "Learners" are the largest group of viewer types who perceive the cultural difference to be large, but they can accept it to a high degree.
>
> H6b: "Learners" are primarily well-educated viewers who know a great deal about foreign cultures.

A small group of participants avoided cultural differences. They either escaped from cultural difference or adhered to Chinese culture. Some kept a neutral attitude. Only a few participants were used to adjusting cultural differences in foreign programs. Therefore:

> H7: The viewer groups of "escapees", "adherents", and "adjusters" are smaller than "learners".

- **Genre-related cultural difference**

As mentioned above, the linguistic content of music television programs is the lowest compared to other television genres. Thus, cultural barrier in music programs with little linguistic content is likely to be lower than other programs in which speech is an important element. Therefore:

> H8: For all viewer types the cultural difference in music programs is the smallest and the most acceptable.

- **Genre-unrelated cultural difference**

As discussed above, language difference might be a major factor in determining cultural barriers and the cultural difference that is the most difficult to accept. Therefore, the following is proposed:

> H9: With language difference viewers are primarily "escapees" and "adherents".

In the personal interviews many participants interpreted their favorite way of life and education method to be ones they had seen in foreign television programs. They strongly showed a preference for learning the new way of life. Therefore:

> H10a: With cultural difference of way of life and education method, viewers are primarily "learners".
> H10b: "Learners" and "adjusters" prefer to learn and adjust their way of life and education methods more than other cultural-difference dimensions.

### 13.2.3 Influential Factors

- **Demographics**

Television viewers are not simply monolithic. Thus, their opinions about cultural difference are the sum total of gender, age, education, and income. These demographic characters are very important control variables in this study, because since the 1980s due to economic development the differences between various demographic groups have increased. In particular, this study tests hypotheses about four demographics:

*Gender*: In traditional gender roles in China, men typically had more power and higher status than women. Men were the breadwinners and heads of the family, whereas women's social roles were narrowly defined within family to producing offspring. However, in the past 30 years gender roles in China have changed from a male domination to a rise in the social position of female in the workplace, society, and education.

Therefore, the opinions given by men and women about cultural difference in foreign television programs might be controversial. They might have different preferences for television genres. For example, men like to watch sports programs and news programs. In contrast, women like to watch television series about family and romance. Furthermore, generally men, who have higher position in society, might have more opportunity to learn out foreign cultures and be more adaptive to cultural differences than women. To test this proposition, this study developed the following two hypotheses:

H11a: Men perceive more cultural differences than women.

H11b: Men accept more cultural differences than women.

*Age*: Due to their different living environments, the generations born before and after 1980 have a large generation gap. The generations born before 1980 adhered their values to mainstream traditional Chinese culture. In contrast, the so-called post-1980s generation is the first after *China's Opening Up Policy*. They have more opportunities to travel, work or study abroad. They have challenged traditional norms and created a mixed value system consisting of both Chinese and Western values.

Therefore, the attitudes of young generations towards cultural difference might be more open and adaptive than those of older generations. Age seems likely to be related to the perception and acceptance of cultural difference:

H12a: Old viewers perceive more cultural difference than young viewers.

H12b: Young viewers accept more cultural difference than old viewers.

*Education*: In China today, education opportunities are unequal for people in rural and urban areas as well as economically impoverished and developed areas. Educational opportunities are also closely related to how much money a family can afford for their children's education. One result is that education levels vary strongly among people in China.

People with a college education have more opportunity of experiencing foreign culture through the internet, television, books, magazines, acquaintances, and so on. They are more likely to get to know something about foreign cultures due to some motivations they might have, such as studying abroad, learning a foreign language, and developing intercultural competence to get a good job. Thus, education level seems likely to exert a positive influence on the reception of cultural difference.

> H13a: Viewers with a high level of education perceive more cultural difference than those with a low level of education.
> H13b: Viewers with a high level of education accept more cultural difference than those with a low level of education.

*Income*: Finally, income difference is large between the rich and poor in China, due to the process of transition from a planned economy to a market economy, which dismantles the equalization of income across sectors, regions and households. People with high incomes might have more opportunities to come into contact with foreign cultures and their minds might be more open and adaptive to them; on the other hand, people with low incomes have to concentrate on fundamental human needs and their outlook is restricted in daily life. Thus, the acceptance degree of cultural difference is predictably higher among the middle and upper social classes than the low class. To test this proposition, this study developed the following two hypotheses:

> H14a: Viewers with high incomes perceive more cultural difference than those with low income.
> H14b: Viewers with high incomes accept more cultural difference than those with low income.

- **Knowledge about foreign culture**

In the communication process of television programs, the program producer encodes and sends the messages they want to transmit to viewers through the television program. The viewer receives and decodes these messages according to their own culture's framework. In this process, whether encoding and decoding is successful depends on how much cultural knowledge, background, and experiences the television producers and viewers have in common. If shared knowledge is low, cultural difference is large for viewers. If they share a lot of knowledge, cultural difference is low for them.

Viewers' knowledge about foreign culture teaches them how to understand foreign programs. If they have a lot of knowledge about foreign culture, they could take another cultural perspective in order to understand things across cultural boundaries based on their ability to use alternative cultural interpretations. Therefore:

H15a: Viewers who know a lot about foreign culture perceive more cultural difference than others who know little.
H15b: Viewers who know a lot about foreign culture accept more cultural difference than others who know little.

- **Viewing foreign television program**

Watching foreign television programs seems likely to exert a positive influence on the reception of cultural difference. The results of the personal interviews indicated that through watching foreign programs, viewers gradually accumulated knowledge about foreign cultures. Step by step they recognized that there were different cultural systems in play in foreign programs. In order to understand them viewers must have longer-term experiences of watching foreign programs in order to become used to a new way of thinking. Based on their accumulated experience, they felt cultural difference to be getting smaller and smaller. To test this proposition, this study developed the following two hypotheses:

H16a: Viewers who watch a lot of foreign TV programs perceive more cultural difference than those who watch little.
H16b: Viewers who watch a lot of foreign TV programs accept more cultural difference than those who watch little.

- **Cultural proximity**

Television programs are more than just a commodity; they are also cultural goods as they say something that must be understood in a cultural context. They are culturally "read" and interpreted by television viewers. Therefore, viewers, in general, have a stronger preference and desire for television content originating from their own culture or a similar culture that reflects a person's own language, cultural background, history, and values. The larger the cultural distance between the exporting and importing markets, the more difficult they are to accept. Thus, the following is proposed:

H17: Viewers like TV programs from culturally proximate countries (Asia) more than culturally distant countries (America, Europe, and other countries).

Although many studies indicated that a large cultural distance between the origin and destination country often caused a lower acceptance in the destination country, the results of the interviews in this study indicated that programs from the United States seem more popular than programs from Japan and Korea, which have similar cultures to China. To test this proposition, this study developed the following hypotheses:

H18a: Viewers who like TV programs from culturally proximate countries (Asia) perceive less cultural difference than those who do not.

H18b: Viewers who like TV programs from culturally proximate countries (Asia) accept less cultural difference than those who do not.

H19a: Viewers who like TV programs from culturally distant countries (America, Europe, and other non-Asian countries) perceive more cultural difference than those who do not.

H19b: Viewers who like TV programs from culturally distant countries (America, Europe, and other non-Asian countries) accept more cultural difference than those who do not.

- **Regional difference**

*Regional difference*: In large geographical regions in China cultures, degree of economic development, and openness to the outside world vary greatly between cities, provinces, and regions. In the east costal region, people have a higher education level and income. They also have more opportunities to use media as information sources to learn about the outside world and to contact people from other cultures in college and the workplace, or friends. In contrast, in some remote areas in central and west China, people do not have any contact with the outside world due to lack of a telecommunication infrastructure. Thus, in general people in economically developed and more open cities might have more open minds towards cultural difference than people in economically impoverished and conservative cities.

*Regional convergence*: Attributed to telecommunication advances, regional cultural differences are less pronounced and local culture may play a diminished role in media consumption.

H20a: Viewers from more open cities perceive more cultural differences than those from more conservative cities.

H20b: Viewers from more open cities accept more cultural differences than those from more conservative cities.

# 14 Results I: Cultural-Difference Dimensions

In the first part of the results, first the perception and acceptance indices *between* ten cultural-difference dimension sets (incl. six genre-related and four genre-unrelated cultural dimension sets) are compared with one another. Then, both indices *among* each cultural-difference dimension set are compared with each other.

As introduced above in the survey, there are a total of 42 cultural-difference dimensions and each one has both perception and acceptance indices which are measured on a 5-point-scale. The perception index (PI) varies between "1 = very small" to "5 = very large" and acceptance index (AI) varies between "1 = totally acceptable" to "5 = not acceptable".

The results of the reliability analysis indicate that the scales of all 42 items through listwise deletion are highly reliable and their variance is statistically significant. Cronbach's $\alpha$ for the perception index of 42 items is .94 ($N = 316$; $F = 20.26$, $p < .001$) and Cronbach's $\alpha$ for the acceptance index of 42 items is .95 ($N = 290$; $F = 17.73$, $p < .001$).

The item analysis was conducted in order to examine the statistical significance of each cultural-difference dimension: (See Appendix 6: Item Analysis)

- First, the missing value is within an acceptable range. For the perception index, the missing value of 36 items is less than 4%, and only 6 items are between 4.22% and 6.89%. For the acceptance index, the missing value of 35 items is less than 5%, and only 7 items are between 5.11% and 8.67%.

- Second, the mean score of each item is also within an acceptable range. The perception indices of each cultural difference item vary between 3.13 and 4.04. The mean score of all perception indices (42 items) is 3.54. The acceptance indices of each cultural difference item vary between 2.01 and 2.76. The mean score of all acceptance indices (42 items) is 2.27.

- Third, the standard deviation of mean score (PI: $1.05 < SD < 1.27$; AI $.95 < SD < 1.26$) and standard error (PI: $.5 < SE < .6$; AI $.5 < SE < .6$) is within an acceptable range.

- On the whole, all of the 42 items for both the perception index and acceptance index are statistically significant.

To make a comparison of various difference dimensions possible, one question must be answered first: Where are the cut-off points of the different degrees of perception and acceptance? Precisely, how should the high, middle, and low levels of both indices be defined?

In this study the most rational way to define the levels of the indices should be according to the evaluation of the participants in the survey. The arbitrary definition of levels without concerning the opinions of participants is not convenient. Each participant has a *total perception index (TPI)*, which is the mean score of the perception indices of all 42 cultural difference items, and a *total acceptance index (TAI)*, which is the mean score of the acceptance indices of all 42 cultural difference items. All TPIs and TAIs are arranged from low to high. The total size of the sample is 450. The range of both indices of the first third of participants (N = 150) is defined as *Level a*, for the second third (N = 150) it is *Level b*, and the last (N = 150) *Level c*. (See Table 21)

Table 21 demonstrates the ranges of the three levels of the perception and acceptance indices as well as the distance between both indices. Each level has a different shade of gray as a background color. For example, the perception index between 1.61 and 3.29 is classified as *Level a* and shown in white; the perception index between 3.30 and 3.87 is classified as *Level b* and shown in light gray; and between 3.88 and 4.95 is classified as *Level c* and shown in dark gray. A PI in white means that a cultural-difference dimension is perceived to be small, while the perception index in dark gray means it is perceived to be large. An AI in white means a cultural-difference dimension is easy to accept; dark gray means it is difficult to accept.

*Table 21: Levels of Perception and Acceptance Indices*

| | Level a | | Level b | | Level c | |
|---|---|---|---|---|---|---|
| Perception | PI - a | 1.61 - 3.29 | PI - b | 3.30 - 3.87 | PI - c | 3.88 - 4.95 |
| Acceptance | AI - a | 1.00 - 1.99 | AI - b | 2.00 - 2.57 | AI - c | 2.58 - 4.28 |
| Distance | DI - a | - .80 - .74 | DI - b | .75 - 1.71 | DI - c | 1.72 - 3.87 |

Source: Own illustration.

## 14.1 Comparison between Cultural-Difference Dimension Sets

In this section, a comparison is made *between* ten sets of cultural-difference dimensions, which include six genre-related sets (movies & TV series, sports programs, documentaries, entertainment shows, news programs, and music programs) and four genre-unrelated sets (language, values, way of life, and education method). Each set includes two to seven particular dimensions.

Both indices of each set are the mean score of both indices of all particular dimensions which the set includes:

$PI_{Set1} = (PI_{Set1\text{-}1} + PI_{Set1\text{-}2} + \ldots PI_{Set1\text{-}n}) / n$

$AI_{Set1} = (AI_{Set1\text{-}1} + AI_{Set1\text{-}2} + \ldots AI_{Set1\text{-}n}) / n$

For example, the perception index of sports programs (M = 3.32, SD = .89) is the mean score of the perception index of all four cultural-difference dimensions which sports programs include: moderation/commentary (M = 3.34, SD = 1.13), sports culture (M = 3.28, SD = 1.13), commercialization (M = 3.51, SD = 1.13), sport types (M = 3.17, SD = 1.27). Both indices of all ten cultural-difference sets are demonstrated and compared with each other in Table 22.

*Table 22: Perception/Acceptance Index, and Distance (Main-Catalogue)*

| | Perception Index | | | Acceptance Index | | | Distance |
|---|---|---|---|---|---|---|---|
| | M | SD | N | M | SD | N | M |
| Entertainment show | 3.64 | 1.05 | 441 | 2.37 | 1.00 | 441 | 1.27 |
| Music program | 3.53 | 1.06 | 439 | 2.04 | .96 | 434 | 1.49 |
| Movie & TV series | 3.48 | .73 | 449 | 2.15 | .82 | 445 | 1.33 |
| Sports program | 3.32 | .89 | 442 | 2.30 | .92 | 435 | 1.02 |
| Documentary | 3.32 | .90 | 445 | 2.16 | .84 | 440 | 1.16 |
| News program | 3.31 | .87 | 435 | 2.29 | .87 | 434 | 1.03 |
| Education method | 3.93 | 1.00 | 445 | 2.15 | .94 | 445 | 1.78 |
| Way of life | 3.87 | 1.00 | 445 | 2.21 | .98 | 445 | 1.67 |
| Language | 3.52 | .92 | 448 | 2.71 | 1.02 | 446 | .81 |
| Value | 3.48 | .74 | 450 | 2.33 | .71 | 450 | 1.15 |
| Total | 3.54 | .63 | 450 | 2.27 | .64 | 450 | 1.27 |

Note: ☐ Level a (low); ▒ Level b (medium); ■ Level c (high).

Source: Own illustration.

- **Cultural difference is large, but acceptable**

H1a assumed that television viewers perceive cultural differences to be large. H1b assumed that viewers mostly accept cultural difference. Both hypotheses are examined through the range of the perception index and acceptance index of ten cultural-difference sets.

Table 20 demonstrates both indices of all ten cultural-difference sets in the main catalogue. The perception indices of all ten sets vary between "3 = normal" and “4 = large”. And the total perception index is M = 3.54 (SD = .63). This indicates that cultural difference could be perceived to be large, but not very large. The acceptance indices of all ten sets vary between “2 = acceptable” and “3 = normal”. And the total acceptance index is M = 2.27 (SD = .64). This indicates that cultural difference could be mostly accepted, but not totally.

=> H1a and H1b were supported by all ten cultural-difference dimension sets in the main catalogue.

- **Genre-related cultural difference**

PIs and AIs of all six genre-related sets stay at the middle level (light gray). Among them *entertainment shows* have the largest PI (M = 3.64, SD = 1.05) and largest AI (M = 2.37, SD = 1.00). This indicates that cultural differences in foreign entertainment shows are the largest and the most difficult television genre to accept. (See Table 22)

The perception index of *music programs* (M = 3.53, SD = 1.06) and *movies & TV series* (M = 3.48, SD = .73) are almost equal. They are lower than entertainment programs, but higher than the other three genres – sports programs, documentaries, and news programs, whose perception indices are almost at the same level of 3.32. The acceptance index of music programs (M = 2.04, SD = .96) is the lowest. This indicates that foreign music programs are the easiest television genre to accept.

H2a assumed that television genres (music programs) with little linguistic content have less of a cultural difference than those in which speech is an important element (news programs). Although music programs contain the lowest linguistic content, their perception index (M = 3.53, SD = 1.06) is obviously higher than for the other four genres – movies & TV series, sports programs, documentaries, and news programs. The news programs in which speech is an important element have the lowest perception index (M = 3.31, SD = .87).

=> H2a was not supported.

H2b assumed that television genres (music programs) with little linguistic content are easier to accept than those in which speech is an important element (news programs). The acceptance index of music programs (M = 2.04, SD = .96) is obviously lower than that of the other five genres. This indicates that music programs which contain low linguistic content are the easiest foreign television genre to accept. The acceptance index of entertainment programs (M = 2.37, SD = 1.00) and news (M = 2.29, SD = .87) are higher than for music programs. This indicates that both genres which contain a high linguistic content are more difficult to accept than music programs.

=> H2b was supported.

- **Genre-unrelated cultural difference**

*(1) Language*

H3a assumed that language difference is the most difficult cultural difference to accept. The results in Table 22 very clearly demonstrate that language difference is the most difficult cultural difference to accept, since its acceptance index is higher (M = 2.71, SD = 1.02) than those of the other three genre-unrelated dimensions – education method (M = 2.15, SD = .94), way of life (M = 2.21, SD = .98), and values (M = 2.33, SD = .71). The distance between the AI of language and the AI of values is .38. Language difference is the only cultural-difference dimension whose acceptance index is at *Level c* – the highest level.

=> H3a was supported very strongly.

H3b assumed that viewers prefer watching foreign television programs with dubbing than those in the original language with subtitles. In the survey the participants were asked to answer the question "which form of language translation in foreign television programs would you rather like?" There are four answers: (1) dubbing; (2) original language with subtitles; (3) original language without subtitles; (4) I do not care.

The results indicate that nearly three quarters participants (73.7%) like to watch foreign programs in the original language with subtitles. 16.4% viewers like to watch them with dubbing. The other 6% do not care of translation forms. Only 3.9% viewers like the original language without subtitles. (See Figure 16) This indicates that most Chinese viewers prefer to watch foreign programs in the original language with or without subtitles. Only a small portion of them like dubbing.

=> H3b was not supported.

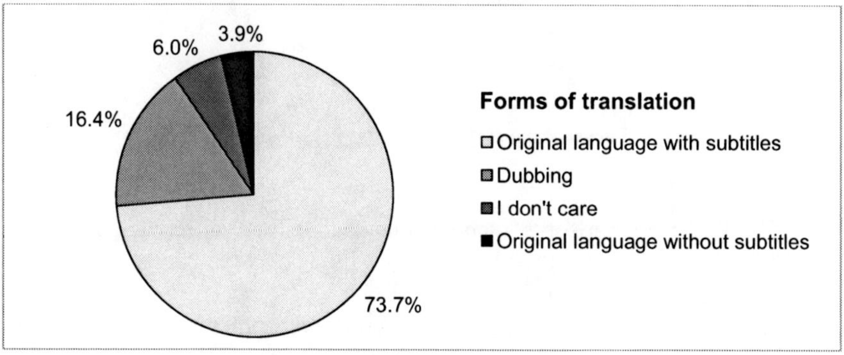


*Figure 16: Forms of Language Translation*

Source: Own illustration.

The reason for this is likely to be, as some participants in the personal interviews explained, that dubbing sounded strange and awkward, whereas the original voices sounded authentic, natural, and moving. They stated that Chinese dubbing could not match foreign movies very well, the feeling and emotion of the original voices were more or less changed or lost in dubbing, and sometimes dubbing was confusing for viewers.

*(2) Education method*

H4a assumed that the distance between the perception index and the acceptance index of education method is large. Table 22 demonstrates that the PI (M = 3.93, SD = 1.00) of education method is the highest and its AI (M = 2.15, SD = .94) is the lowest. This result very clearly indicates that education method is the largest cultural difference, but it is the easiest to accept. The distance between the PI and AI reached 1.78, which is the largest distance compared to all other nine cultural-difference sets. (See Table 22) This result corresponds to the opinions expressed by the interviewees. They thought that the education methods they had seen in foreign television programs were very different, because they seemed more open and freer than those used in China. They would like to educate their children in the foreign way that had seen.

=> H4a was supported.

*(3) Way of life*

H4b assumed that the distance between the perception and acceptance indices of way of life is large. Like in the case of education method, way of life is also perceived to be very large (M = 3.87, SD = 1.00), but it is very easy to accept (M = 2.21, SD = .98). The distance between the PI and AI reached 1.67, which is the second largest distance behind education method. This result could be explained by the preference for a foreign way of life expressed by many Chinese television viewers. (See Table 22)

=> Therefore, H4b was supported.

## 14.2 Comparison among Cultural-Difference Dimension Sets

In this section, a comparison is made *among* each cultural-difference dimension set. Table 23 demonstrates the PI and AI as well as distance between both indices of all 42 particular dimensions.

In Table 23 the background colors of the 42 perception indices are mixture of white (low level), light gray (middle level), and dark gray (high level). The perception indices vary between 3.13 and 4.04. This indicates that these cultural differences are generally large, but not very large. 37 acceptance indices remain at a middle level and only five dimensions of language at a high level. The acceptance indices vary between 2.01 and 2.76. This indicates that generally these cultural differences are mostly accepted, but not totally.

=> H1a and H1b were also supported for all 42 particular dimensions in the sub-catalogue.

*Table 23: Perception/Acceptance Index, and Distance (Sub-Catalogue)*

| | | L. | Perception Index | | | Acceptance Index | | | Dis. |
|---|---|---|---|---|---|---|---|---|---|
| | | | M | SD | N | M | SD | N | M |
| Set 01 | Movie & TV series | | | | | | | | |
| 01 - 1 | Appearance of character | A | 3.34 | 1.08 | 446 | 2.08 | 1.06 | 442 | 1.26 |
| 01 - 2 | Personality of character | A | 3.75 | 1.08 | 444 | 2.19 | 1.08 | 441 | 1.55 |
| 01 - 3 | Music style | A | 3.23 | 1.21 | 439 | 2.07 | 1.09 | 437 | 1.16 |
| 01 - 4 | Theme of drama plot | A | 3.60 | 1.13 | 441 | 2.13 | 1.05 | 438 | 1.46 |
| 01 - 5 | Proximity to daily life of d.p. | A | 3.22 | 1.11 | 435 | 2.29 | 1.16 | 430 | .93 |
| 01 - 6 | Fantasy of drama plot | A | 3.81 | 1.15 | 419 | 2.10 | 1.15 | 416 | 1.71 |
| Set 02 | Sports program | | | | | | | | |
| 02 - 1 | Moderation/commentary | B | 3.34 | 1.13 | 440 | 2.30 | 1.06 | 431 | 1.04 |
| 02 - 2 | Sports culture | B | 3.28 | 1.13 | 433 | 2.28 | 1.10 | 427 | 1.01 |
| 02 - 3 | Commercialization | B | 3.51 | 1.13 | 422 | 2.48 | 1.19 | 413 | 1.03 |
| 02 - 4 | Sports types | A | 3.17 | 1.27 | 434 | 2.12 | 1.07 | 431 | 1.05 |
| Set 03 | Documentary | | | | | | | | |
| 03 - 1 | Objectivity vs. subjectivity | B | 3.27 | 1.15 | 443 | 2.18 | .98 | 438 | 1.09 |
| 03 - 2 | Explicit vs. implicit | B | 3.26 | 1.10 | 441 | 2.15 | .95 | 437 | 1.11 |
| 03 - 3 | Starting point/perspective | B | 3.45 | 1.15 | 442 | 2.19 | 1.01 | 437 | 1.25 |
| 03 - 4 | Theme | A | 3.32 | 1.16 | 437 | 2.12 | 1.04 | 432 | 1.21 |
| Set 04 | Entertainment show | | | | | | | | |
| 04 - 1 | Speaking style of presenter | A | 3.76 | 1.15 | 435 | 2.39 | 1.12 | 436 | 1.38 |
| 04 - 2 | Content of entertainment | A | 3.52 | 1.17 | 440 | 2.34 | 1.06 | 440 | 1.18 |
| Set 05 | News | | | | | | | | |

| | | | | | | | | | |
|---|---|---|---|---|---|---|---|---|---|
| 05 - 1 | Appearence of presenter | A | 3.13 | 1.14 | 434 | 2.22 | 1.02 | 428 | .91 |
| 05 - 2 | Personality of presenter | A | 3.35 | 1.06 | 433 | 2.29 | 1.01 | 430 | 1.06 |
| 05 - 3 | Objectivity vs. subjectivity | B | 3.25 | 1.15 | 431 | 2.33 | 1.03 | 420 | .92 |
| 05 - 4 | Starting point/perspective | B | 3.30 | 1.10 | 422 | 2.34 | 1.02 | 416 | .96 |
| 05 - 5 | Diversity of news theme | B | 3.36 | 1.13 | 424 | 2.29 | 1.03 | 415 | 1.08 |
| 05 - 6 | Sensitivity of news theme | B | 3.49 | 1.17 | 422 | 2.31 | 1.07 | 411 | 1.18 |
| Set 06 | Music program | | | | | | | | |
| 06 - 1 | Music style | A | 3.56 | 1.17 | 439 | 2.07 | 1.03 | 433 | 1.49 |
| 06 - 2 | Music quality | A | 3.51 | 1.16 | 437 | 2.01 | 1.01 | 433 | 1.49 |
| Set 07 | Language | | | | | | | | |
| 07 - 1 | Voice of dubbing | A | 3.55 | 1.21 | 447 | 2.71 | 1.20 | 445 | .84 |
| 07 - 2 | Emotion of dubbing | B | 3.54 | 1.12 | 444 | 2.67 | 1.17 | 443 | .87 |
| 07 - 3 | Coordination of dubbling | B | 3.42 | 1.12 | 445 | 2.72 | 1.18 | 444 | .70 |
| 07 - 4 | Acccuracy of translation | B | 3.49 | 1.08 | 443 | 2.76 | 1.18 | 438 | .73 |
| 07 - 5 | Humor in translation | B | 3.65 | 1.14 | 444 | 2.71 | 1.26 | 444 | .94 |
| Set 08 | Value | | | | | | | | |
| 08 - 1 | Individualism | C | 3.65 | 1.09 | 446 | 2.27 | .99 | 442 | 1.38 |
| 08 - 2 | Religion | C | 3.74 | 1.12 | 446 | 2.51 | 1.12 | 441 | 1.23 |
| 08 - 3 | Morality | C | 3.45 | 1.08 | 443 | 2.39 | 1.06 | 441 | 1.06 |
| 08 - 4 | Family | C | 3.42 | 1.11 | 446 | 2.33 | 1.03 | 441 | 1.09 |
| 08 - 5 | Liberty/democracy/equality | C | 3.39 | 1.24 | 445 | 2.25 | 1.06 | 444 | 1.14 |
| 08 - 6 | Social status/competition | C | 3.57 | 1.16 | 447 | 2.35 | 1.02 | 443 | 1.22 |
| 08 - 7 | Patriotism | C | 3.13 | 1.23 | 444 | 2.26 | 1.07 | 440 | .87 |
| Set 09 | Way of life | | | | | | | | |
| 09 - 1 | Habits/customs/rituals | B | 3.91 | 1.05 | 445 | 2.22 | 1.05 | 444 | 1.70 |
| 09 - 2 | Living environment | A | 3.83 | 1.09 | 442 | 2.19 | 1.03 | 441 | 1.64 |
| Set 10 | Education method | | | | | | | | |
| 10 - 1 | Freedom/independence | B | 3.90 | 1.18 | 445 | 2.10 | 1.03 | 445 | 1.79 |
| 10 - 2 | Imagination/creativity | B | 4.04 | 1.15 | 443 | 2.10 | 1.09 | 444 | 1.94 |

| | | | | | | | | | |
|---|---|---|---|---|---|---|---|---|---|
| 10 - 3 | Equality between young&old | B | 3.84 | 1.25 | 444 | 2.19 | 1.08 | 443 | 1.65 |
| 10 - 4 | Demand on children | B | 3.95 | 1.14 | 444 | 2.20 | 1.11 | 444 | 1.75 |

Note: [white box] Level a (low); [light grey box] Level b (medium); [dark grey box] Level c (high).
L. = Cultural level; Dis. = Distance between PI and AI.
Source: Own illustration.

- **Genre-related cultural difference**

*Movie & television series*: Among six dimensions the perception index of "fantasy of drama plot" is the largest (M = 3.81, SD = 1.15). The PIs of "proximity to daily life of drama plot" (M = 3.22, SD = 1.11) and "music style" (M = 3.23, SD = 1.21) are the lowest at Level A. (See Table 23)

The acceptance indices of all six dimensions do not have large differences and they stay at medium Level B. The most unacceptable difference is "proximity to daily life of drama plot" (M = 2.29, SD = 1.16) and the most acceptable is "music style" (M = 2.07, SD = 1.09) and "appearance of character" (M = 2.08, SD = 1.06). An analysis of the distance index reveals that the DI of "fantasy of drama plot" (M = 1.71) is obviously larger than for the other six dimensions. This means that although the cultural difference of imagination is large, Chinese viewers accept it very well or even favor it a lot.

*Sports programs*: Among four particular dimensions "commercialization" is the largest (PI: M = 3.51, SD = 1.13) and the most difficult to accept (AI: M = 2.48, SD = 1.19). "Sports types" is the smallest (PI: M = 3.17, SD = 1.27) and the easiest dimension to accept (AI: M = 2.12, SD = 1.07). (See Table 23)

*Documentary*: Both indices of all four dimensions are almost equal.

*Entertainment show*: The difference of "speaking style of presenter" is relatively large (M = 3.76, SD = 1.15).

*News program*: "Sensitivity of news theme" is the largest difference.

*Music program*: The perception indices of both dimensions are relatively large, but relatively easy to accept. (See Table 23)

The perception indices of the two dimensions of music programs with low linguistic content are larger than all six dimensions of news programs in which speech is an important element.

=> H2a was not supported for any of dimensions in the sub-catalogue.

Furthermore, the acceptance indices of the two dimensions of music programs are lower than all six dimensions of news programs.

=> H2b was supported for all dimensions in the sub-catalogue.

- **Genre-unrelated cultural difference**

*Language*: All five language dimensions have the highest acceptance index among all 42 dimensions. All these five acceptance indices stay at a high level. This indicates that, on the whole, language difference is the most difficult cultural difference to accept. Among them "accuracy of translation" has the largest acceptance index (M = 2.67, SD = 1.18) than other four. (See Table 23)

=> H3a was also supported for all five language difference dimensions in the sub-catalogue.

*Value*: Generally, all seven value differences are large. Among them difference in "religion" is the largest (M = 3.74, SD = 1.12) and the most unacceptable (M = 2.51, SD = 1.12). "Patriotism" is the smallest value difference (M = 3.13, SD = 1.23) and the most acceptable (M = 2.26, SD = 1.07). The other four value differences are almost equal. The distance index of "individualism" (M = 1.38) is the largest, as some interviewees stated that they looked forward to an independent life like the one they had seen in foreign television programs. (See Table 23)

*Education method*: The perception index of "imagination/creativity" (P = 4.04, SD = 1.15) is the largest of all 42 dimensions and it is also the only one larger than 4. However, participants accept them mostly (M = 2.10, SD = 1.09). Thus, its distance index is also the largest (M = 1.94). Besides "imagination/creativity" the other three dimensions of education method also seem very large, but they are easy to accept. (See Table 23)

=> H4a is also supported for all four dimensions in the sub-catalogue.

*Way of life*: Chinese viewers also favor "habits/customs/rituals" (PI: M = 3.91, SD = 1.05; AI: P = 2.22, SD = 1.05) and "living environment" (PI: M = 3.83, SD = 1.09; AI: P = 2.19, SD = 1.03) that were presented on foreign television. Although they perceive both differences to be large, they could mostly accept them. (See Table 23)

=> H4b was also supported for the two dimensions in the sub-catalogue.

- **Cultural level**

Table 23 also shows the cultural levels of each cultural-difference dimension. *Level A* is the surface level, which is explicitly clear and highly visible, which could be, for example, the appearance of characters in movies and the living environment. *Level B* is the middle level, which is at least partially visible, and relates to aspects like habits, customs, and rituals. *Level C* is the deepest level of culture and its essence, which is largely invisible and unconscious, such as the case for values.

H5a assumed that cultural difference at a deep level is perceived to be larger than cultural difference at surface level. H5b assumed that cultural difference at a deep level is less acceptable than cultural difference at surface level. To examine these two hypotheses a regression analysis was conducted where level A was coded as 1, level B as 2, and level C as 3.

The results of the regression analysis indicate that cultural level (M = 1.79, SD = .74, N = 42) is neither correlated to the perception index (M = 3.51, SD = .24, N = 42) nor to the acceptance index (M = 2.29, SD = .19, N = 42). This reveals that cultural difference at a deep level is not necessarily larger than at surface level. And cultural difference at a deep level is not necessarily more difficult to accept than it on a superficial level.

=> H5a and H5b were not supported.

# 15 Results II: Television Viewer Types

The Chapter 14 demonstrated the results of comparison of the perception and acceptance indices of various cultural-difference dimensions. However, they could not indicate the similarities and dissimilarities between viewers. Therefore, in Chapter 15 a cluster analysis is used to examine how similar the opinions are about cultural differences between viewers.

In the results of the personal interviews, all interviewees were divided into five different viewer types according to perception and acceptance degree: escapees, learners, adherents, adjuster, and neutralists. In the survey all participants are also divided into several viewer types according to their perception index and acceptance index.

## 15.1 Large Cultural Difference, But Mostly Acceptable

The results of the cluster center analysis are presented in a scatter diagram (see Figure 17). Each viewer/dot has a total perception index (TPI) and a total acceptance index (TAI). Both represent the mean score of all 42 cultural-difference dimensions.

In the scatter diagram the horizontal axis is the perception index from "1 = very small" to "5 = very large" ($F = 243.37$, $p < .001$) and the vertical axis is the acceptance index from "1 = totally acceptable" to "5 = not acceptable" ($F = 448.85$, $p < .001$). The area between the horizontal and vertical axes is divided into four particular areas. The dots in the upper-right *area A* have a large PI and a large AI. The dots in the lower-right *area B* have a large PI but a low AI. Dots in the upper-left *area C* have a low PI but a high AI. Finally, in the lower-left *area D* they have a low PI and a low AI. The dots in the middle *area E* have a middle PI and a middle AI.

It is important to point out that the middle point of the diagram (PI = 3, AI = 3) is artificially defined, which is an ideal neutral point for distinguishing five viewer types. In the results of the cluster center analysis, the actual middle point may vary near to this ideal middle point.

The scatter diagram (Figure 17) very clearly presents that most dots stay in the lower-right area B; in other words, under a diagonal between two extreme points (PI = 1, AI = 1) and (PI = 5, AI = 5). Based on the results of the cluster analysis with a listwise deletion of missing values all participants are divided into five clusters: escapees, learners, extreme learners, adjusters, and neutralists.

Table 24 describes each cluster. The reasons why the samples are divided into five clusters are the following: (1) six clusters seem too complicated to describe the similarity between participants; (2) four clusters are not detailed enough to interpret the difference between participants.

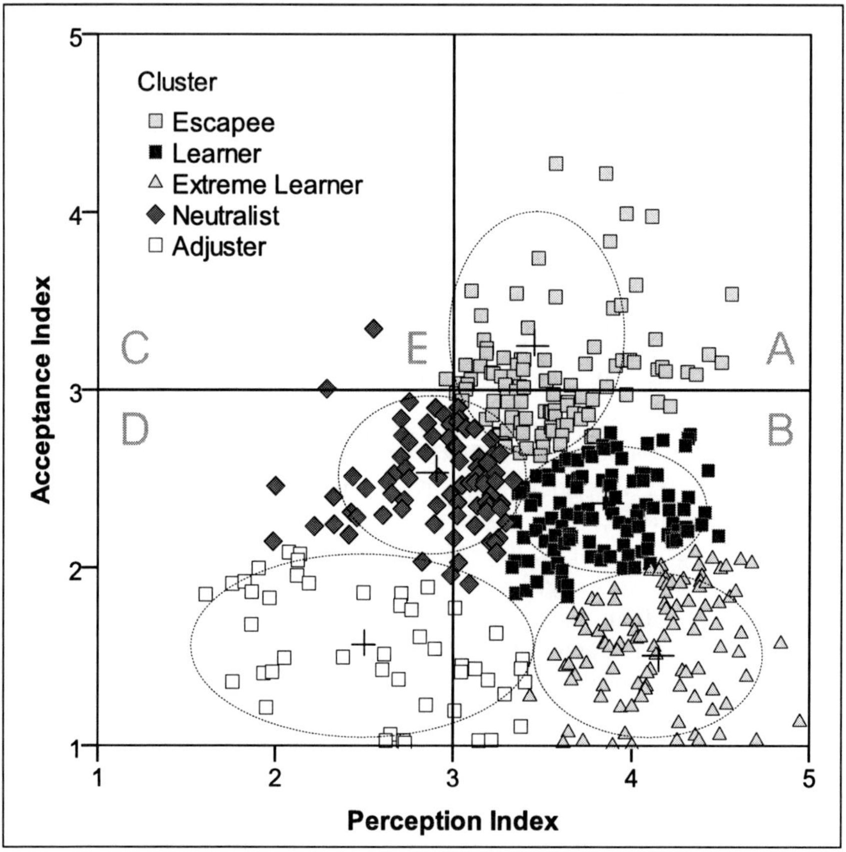


*Figure 17: Clusters of Television Viewers (Scatter Diagram)*
Note: Perception index (1 = very small, 5 = very large);
acceptance index (1 = totally acceptable, 5 = not acceptable)
Source: Own illustration.

*(1&2) Learners and extreme learners*: A large group of 48.0% of participants (N = 216) believe that cultural differences are large, but they mostly have an open mind to accepting them. This large group consists of "learners" (N = 113) and "extreme learners" (N = 103). The difference between both groups is that "ex-

treme learners" (AI: M = 1.57) has a more acceptable attitude towards cultural difference than "learners" (AI: M = 2.30). The domination of both "learner" groups indicates that a large proportion of Chinese television viewers are curious about different cultures and they would like to learn something more about them.

=> H6a which assumed that "(extreme) learner" is the largest group of viewer types was strongly supported.

*(3) Escapees*: Just 24.0% of participants think that cultural differences are large and they could not accept them. This cluster "escapees" is the only one whose center stays in area A (PI: M = 3.55; AI: M = 3.07). This cluster stays above all other four clusters. This means that escapees could not accept cultural difference the most. Escapees (N = 108) are also a large part of sample, but there are fewer of them than learners (N = 216). In the diagram the escapees seem to show a trend of moving towards becoming learners, because the acceptance index of its cluster center is 3.07. This is just a little larger than the neutral level (AI = 3).

*(4) Adjusters*: Their cluster center is located in area D, which has the lowest perception index (M = 2.58) and the lowest acceptance index (M = 1.54). Adjusters have the smallest cluster. Only 10.4% of participants (N = 47) adjusted their own culture towards foreign cultures. Unconsciously, they perceive few cultural differences in foreign television programs and could mostly accept them.

*(5) Neutralists*: The cluster center of "neutralists" (PI: M = 2.91; AI: M = 2.49) is near to the center of the diagram. The results indicate that 17.6% of participants (N = 79) do not pay much attention to cultural difference and their attitude towards cultural difference is neutral. They perceive cultural difference as neither large nor small, and they neither accept it nor not accept it.

*Table 24: Cluster Description*

| **Cluster** | **Perception** | **Acceptance** | **N** | **Percentage** |
|---|---|---|---|---|
| Escapees | 3.55 | 3.07 | 108 | 24.0% |
| Learners | 3.83 | 2.30 | 113 | 25.1% |
| Extreme Learners | 4.14 | 1.57 | 103 | 22.9% |
| Adjusters | 2.58 | 1.54 | 47 | 10.4% |
| Neutralists | 2.91 | 2.49 | 79 | 17.6% |
| Total | | | 450 | 100% |

Source: Own illustration.

There is no cluster in area C. This indicates that the viewer type "adherents" who perceive there to be little cultural difference but could not accept it and adhere to their own culture, is not a typical viewer type in this study.

To investigate each cluster in more detail a Pearson's chi-squared test was conducted to examine the sample distribution in five clusters. The results reveal that education level (Chi-Square = 31.48, df = 8, $p < .001$), knowledge about foreign culture (Chi-Square = 49.35, df = 8, $p < .001$), and city of residence (Chi-Square = 30.12, df = 16, $p < .05$) have a great impact on clusters. (See Table 25)

*Table 25: Sample Distribution in five Clusters*

| | | Clusters (N) | | | | | Σ |
|---|---|---|---|---|---|---|---|
| | | Escapee | Learner | E. learn. | Adjuster | Neutral. | |
| Gender | Male | 40 | 62 | 45 | 22 | 33 | 202 |
| | Female | 65 | 49 | 57 | 25 | 43 | 239 |
| Age | 20 - 29 | 90 | 95 | 85 | 38 | 67 | 375 |
| | 30 - 49 | 16 | 17 | 17 | 9 | 8 | 67 |
| Education | Pri./mid. school | 36 | 15 | 12 | 12 | 21 | 96 |
| | Bachelor | 64 | 89 | 74 | 34 | 48 | 309 |
| | Above master | 5 | 8 | 16 | 1 | 6 | 36 |
| Income | ¥ 0 - 999 | 28 | 29 | 24 | 9 | 15 | 105 |
| | ¥ 1000 - 4999 | 65 | 66 | 54 | 30 | 49 | 264 |
| | Over ¥ 5000 | 7 | 14 | 19 | 7 | 8 | 55 |
| Knowledge | Much | 7 | 14 | 26 | 8 | 8 | 63 |
| | Normal | 50 | 70 | 65 | 35 | 43 | 263 |
| | Little | 44 | 27 | 9 | 4 | 24 | 108 |
| City | Beijing | 13 | 23 | 27 | 18 | 15 | 96 |
| | Shanghai | 20 | 22 | 19 | 10 | 13 | 84 |
| | Guangzhou | 29 | 19 | 13 | 5 | 21 | 87 |
| | Nanjing | 19 | 24 | 25 | 7 | 9 | 84 |
| | Xi'An | 27 | 25 | 19 | 7 | 21 | 99 |

Note: E. learn. = Extreme learners, Neutral. = Neutralists.

*Education*: The viewers with primary-/middle-school education are primarily "escapees" (N = 36), which is obviously larger than the other four viewer types. The largest group of viewers with a bachelor degree is "learners" (N = 89). And "extreme learners" are primarily viewers with an education level above master's (incl. doctorate) (N = 16). This indicates that viewers with a higher education level are more likely to accept cultural difference than viewers with a lower education level.

*Knowledge*: Among participants, who know "much" about foreign culture, the viewer groups of "learners" (N = 14) and "extreme learners" (N = 26) are the largest. Similarly, among participants, who know a "normal" amount about foreign culture, "learners" (N = 70) and "extreme learners" (N = 65) are the two largest groups. In contrast, the largest viewer group among participants, who know "little", is "escapees" (N = 44).

=> H6b, which assumed that "learners" are primarily the well-educated viewers who have a great deal of knowledge about foreign culture, is supported.

*City of residence*: In Beijing the largest viewer types are "learners" (N = 23) and "extreme learners" (N = 27). Like Beijing, the largest viewer types in Nanjing are also "learners" (N = 24) and "extreme learners" (N = 25). However, the viewer group "escapees" is the largest in Guangzhou (N = 29) and Xi'An (N = 27).

In sum, all participants are divided into five clusters. Most of the participants are concentrated in area B with a high PI and a low AI. A large group of 48% of participants (N = 216) are "(extreme) learners", who believe that cultural differences are large, but they mostly have an open mind towards accepting them. Among them, 103 participants are "extreme learners". Most participants with a high education level and a lot of knowledge about foreign culture are "(extreme) learners", whereas the viewers with a low education level and little knowledge about foreign culture are primarily "escapees". Beijing and Nanjing have a large group of "(extreme) learners", while Guangzhou and Xi'An have a large group of "escapees".

## 15.2 Viewer Types

In the following, two simplified diagrams (Figure 18 and 19) are used to demonstrate all of the clusters in a simply way, which makes the comparison of the ten cultural-differences sets possible. Like for the scatter diagram all of the samples of each dimension set are divided into five clusters. The comparison of the ten dimension sets is interpreted from two perspectives: (1) position of cluster and (2) size of cluster. In Figures 18 and 19 the center of each cluster determines its position the in diagram and in Tables 26 and 27 the number of samples in each cluster determines its size.

### 15.2.1 Viewer Types of Genre-Related Cultural Difference

In Figure 18 and Table 26, six genre-related dimension sets are compared with each other (incl. movies & TV series, sports programs, documentaries, entertainment shows, news, and music programs). Each genre has five cluster centers and thus a total of 30 cluster centers are distributed in four areas in the diagram. The cluster centers, which interpret a same viewer type, belong to the same circle. A total of five circles are used to interpret the five viewer types. (See Figure 18)

- **Cluster position**

*(1) Escapees (area A)*

The circle in area A interprets the viewer type "escapees" which includes the cluster centers of all six genres. In this circle, music programs (M (PI) = 4.46, M (AI) = 3.76, N = 34) and entertainment shows (M (PI) = 4.48, M (AI) = 3.73, N = 69) are overlapping. Both have the largest perception and acceptance indices. This indicates that for "escapees" cultural differences in music programs and entertainment shows are the largest and the most unacceptable. (See Figure 18 and Table 26)

*(2&3) Learners and extreme learners (area B)*

There are two circles in area B: "learners" and "extreme learners".

The "learners" circle only contains cluster centers of four genres. The other two genres sports programs and entertainment shows do not have the viewer type of "learners". Among these four genres the cluster center of music programs (M (PI) = 3.09, M (AI) = 1.46, N = 92) is quite remote compared to the other three. It tends to move to area D. This indicates that for "learners" the cultural difference in music programs is the smallest and the most acceptable. Their attitudes towards music programs are clearly more open than for the other three genres. (See Figure 18 and Table 26)

In the "extreme learners" circle all six cluster centers stay relatively close to each other. Among them, documentaries (M (PI) = 4.25, M (AI) = 1.17, N = 83) is the most acceptable genre. This indicates that "extreme learners" would like to accept cultural difference in foreign documentary more than the other five genres.

*(4) Adjusters (area D)*

In the "adjusters" circle, music programs (M (PI) = 1.56, M (AI) = 1.55, N = 43) are somewhat isolated from the other genres on the left-hand side, whereas the cluster center of movies & TV series (M (PI) = 2.44, M (AI) = 1.50, N = 67) are situated away from the other genres on the right-hand side. This indicates that the "adjuster" viewers could accept both genres at an equal level, but they find cultural difference in music programs to be clearly larger than movies & TV series. (See Figure 18 and Table 26)

*(5) Neutralists (middle of four areas)*

The "neutralists" circle is in the middle of the diagram. All six cluster centers are relatively close to one another. Among them, movies and TV series (M (PI) = 3.05, M (AI) = 3.04, N = 96) is located nearer to the mid-point of the diagram than the rest. This means that "neutralists" find cultural difference in foreign movies & TV series to be more neutral than the other five genres. (See Figure 18 and Table 26)

*(6) Adherents (area C)*

There are only two cluster centers in "adherents" in area C: sports programs and entertainment shows. Although cultural difference in sports programs (M (PI) = 2.81, M (AI) = 4.30, N = 18) seems small for "adherents", their own culture dominate in their mind while watching foreign sports programs. In contrast to sports programs, the culture in foreign entertainment shows (M (PI) = 2.09, M (AI) = 3.13, N = 32) seems very similar, but "adherents" still adhere to own culture slightly. (See Figure 18 and Table 26)

In sum, "escapees" find cultural difference in music programs and entertainment shows is the largest and the most unacceptable. "Learners" find cultural difference in music programs to be the smallest and it represents the most acceptable genre for them. "Extreme learners" accept cultural difference in documentaries the most. For "adjusters" music programs have the smallest cultural difference, whereas movies & TV series have the largest. All six cluster centers of "neutralists" are close to each other. In the "adherents" area, there are only sports programs and entertainment shows. They find cultural difference in both genres to be similar to Chinese culture, but they still want to adhere to Chinese culture.

Among the six genres the attitudes towards music programs are the most controversial. "Escapees" think cultural difference in music programs is the largest and the most unacceptable, whereas "adjusters" think cultural difference in music programs is the smallest and for them is the most acceptable genre.

=> H8, which assumed that cultural difference in music programs with little linguistic content is the smallest and is the most acceptable for all viewer types, is not supported.

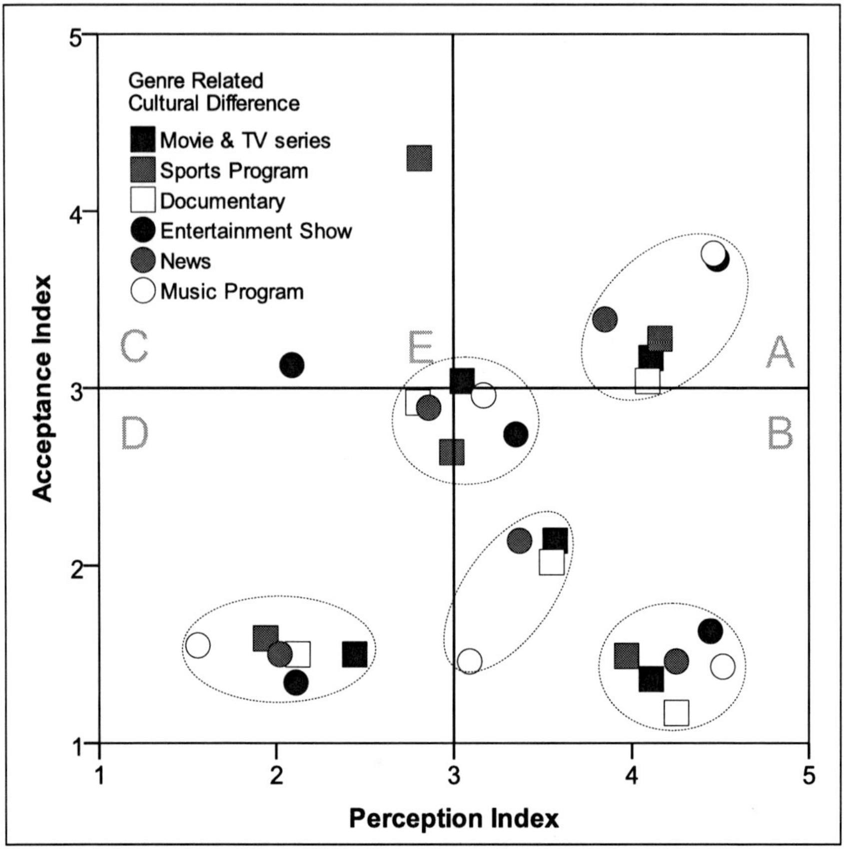


*Figure 18: Comparison of Genre-Related Cultural Difference*
Note: Perception index (1 = very small, 5 = very large);
acceptance index (1 = totally acceptable, 5 = not acceptable)
Source: Own illustration.

- **Cluster size**

*(1) Domination of "(extreme) learners" and "neutralists"*

The results of the cluster size comparison reveal that among the six viewer types "extreme learners" and "neutralists" dominate in the genres sports programs, entertainment shows, news programs, and music programs. (See Table 26)

- *Sports programs*: "extreme learners" (N = 135) and "neutralists" (N = 146) occupy 64.6% of samples.

- *Entertainment shows*: They reach an equal level of 63.9%, of which 150 are "learners" and 130 are "neutralists".

- *Music programs*: They have a lower percentage of 61.1%, which comprises 149 "extreme learners" and 116 "neutralists".

- *News programs*: "extreme learners" (N = 102) and "neutralists" (N = 111) have a lower percentage (49.1%) than the three genres mentioned above.

Different as above four genres:

- *Movies and TV series*: Here "learners" (N = 111) and "extreme learners" (N = 118) dominate. This means that the attitude of a large group of participants towards cultural difference in movies and television series is to keep learning about it.

- *Documentaries*: This also has two viewer types which dominate – "learners" (N = 94) and "neutralists" (N = 110).

Among the six genres the amount of "extreme learners" of entertainment shows (N = 150) and music programs (N = 149) is the largest which occupy one third of the total sample size. Sports programs have the largest number of "neutralists" (N = 146). (See Table 26)

*(2) Subordination of "escapees", "adherents", and "adjusters"*

- For movies & TV series both groups, "escapees" (N = 53) and "adjusters" (N = 67), are also the smallest, comprising 27.0% of the samples. Similarly, "escapees" and "adjusters" are also the smallest viewer groups for documentaries (N = 74 and 79, 34.8%) and music programs (N = 34 and 43, 17.7%). (See Table 26)

- For sports programs, "adherents" (N = 18) and "adjusters" (N = 66) comprised only 19.3% of the total sample size, and also for entertainment programs both viewer types occupied 20.5%.

On the whole, the domination of "(extreme) learners" and "neutralists" indicates the attitudes towards cultural difference of most participants are either to open their mind and learn about cultural difference actively, or to keep a neutral mind and avoid cultural difference. The subordination of "escapees", "adherents", and "adjusters" indicates that only a small group of participants would like either to escapee from foreign culture and adhere to their own indigenous culture, or adjust it to their own indigenous culture.

=> H7, which assumed that the viewer groups of "escapee", "adherent", and "adjuster" are subordinated compared to "learners", was supported by genre-related cultural difference.

*Table 26: Cluster Description of Genre-Related Cultural Difference*

| | | Movie/Se. | | Sports | | Documen. | | Entertain. | | News | | Music | |
|---|---|---|---|---|---|---|---|---|---|---|---|---|---|
| | | | N | | N | | N | | N | | N | | N |
| Escapee | PI | 4.11 | 53 | 4.16 | 70 | 4.09 | 74 | 4.48 | 69 | 3.85 | 76 | 4.46 | 34 |
| | AI | 3.17 | | 3.28 | | 3.04 | | 3.73 | | 3.39 | | 3.76 | |
| Learner | PI | 3.57 | 111 | | | 3.55 | 94 | | | 3.37 | 75 | 3.09 | 92 |
| | AI | 2.14 | | | | 2.02 | | | | 2.14 | | 1.46 | |
| Extreme Learner | PI | 4.11 | 118 | 3.97 | 135 | 4.25 | 83 | 4.44 | 150 | 4.25 | 102 | 4.51 | 149 |
| | AI | 1.36 | | 1.49 | | 1.17 | | 1.63 | | 1.46 | | 1.43 | |
| Adherent | PI | | | 2.81 | 18 | | | 2.09 | 32 | | | | |
| | AI | | | 4.30 | | | | 3.13 | | | | | |
| Adjuster | PI | 2.44 | 67 | 1.94 | 66 | 2.12 | 79 | 2.11 | 58 | 2.02 | 70 | 1.56 | 43 |
| | AI | 1.50 | | 1.59 | | 1.50 | | 1.34 | | 1.50 | | 1.55 | |
| Neutralist | PI | 3.05 | 96 | 2.99 | 146 | 2.80 | 110 | 3.35 | 131 | 2.86 | 111 | 3.17 | 116 |
| | AI | 3.04 | | 2.64 | | 2.92 | | 2.74 | | 2.89 | | 2.96 | |
| Total | N | | 445 | | 435 | | 440 | 440 | | 434 | | 434 | |

Note: Se. = TV series, Documen. = Documentary, Entertain. = Entertainment program.
Source: Own illustration.

## 15.2.2 Viewer Types of Genre-Unrelated Cultural Difference

In Figure 19 and Table 27 four genre-unrelated cultural-difference dimensions are compared with each other. They are language, values, way of life, and education methods. Each dimension has five cluster centers and thus a total of 20 cluster centers distributed across four areas in the diagram. The cluster centers for a certain viewer type belong to the same circle. Unlike genre-related cultural dimensions, the cluster centers of genre-unrelated dimensions in certain circles seem more removed from each other. (See Figure 19)

- **Cluster position**

(1) *Escapees*: In the "escapees" circle in area A, language difference is the largest and most unacceptable cultural difference. Both its perception index (M

= 4.59) and the acceptance index (M = 4.09) are the highest, and thus are clearly higher than for the other three dimensions. In contrast, the value difference is the smallest and the most acceptable. Both its perception index (M = 3.95) and the acceptance index (M = 3.18) are clearly lower than for the others. This indicates that language difference is one kind of cultural difference, which "escapees" want to escapee from the most, wheareas the value difference is what they want to escapee from the least. (See Figure 19 and Table 27)

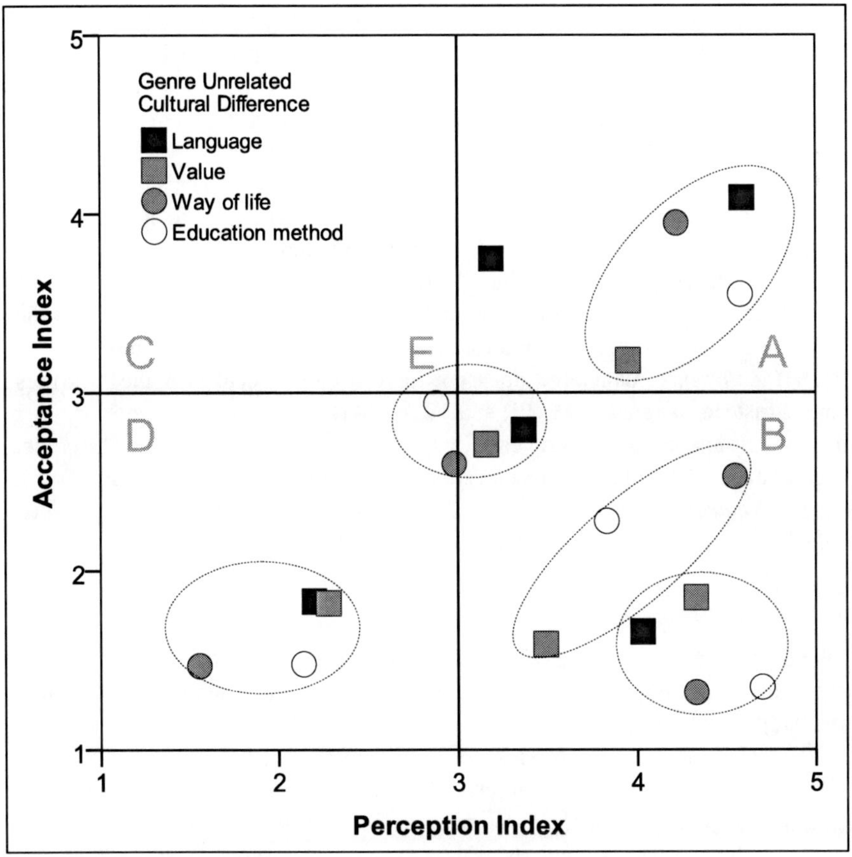


*Figure 19: Comparison of Genre-Unrelated Cultural Difference*
Note: Perception index (1 = very small, 5 = very large);
acceptance index (1 = totally acceptable, 5 = not acceptable)
Source: Own illustration.

(2) *Learners*: In this circle there are only three cluster centers: way of life, education method, and values. They are quite removed from each other. The cluster center for way of life (M (PI) = 4.55, M (AI) = 2.53, N = 102) is situated in the upper-right corner in area B, wheareas the value (M (PI) = 3.49, M (AI) = 1.59, N = 67) is in the lower-left corner in area B. This indicates that "learners" find the cultural difference of way of life to be the largest, and preferred to learn it the least, while value difference is the smallest cultural difference and the one most preferred to be learned. Quite different to "learners", "extreme learners" think the difference between education methods is the largest (M (PI) = 4.70), but they accept it and want to learn about this the most (M (AI) = 1.35). (See Figure 19 and Table 27)

(3) *Adjusters*: In the "adjusters" circle, the perception index of way of life is the smallest (M (PI) = 1.55). The acceptance indices of both way of life (M (AI) = 1.47) and education method (M (AI) = 1.48) are equal. This indicates that "adjusters" preferred to adjust the differences of way of life and education method more han the differences of language and values.

=> H10b, which assumed that "learners" and "adjusters" preferred to learn and adjust way of life and education method rather than other cultural-difference dimensions, was partly supported for "extreme learners" and "adjusters".

(4) *Adherents*: Language does not have a cluster center in the "learners" circle. Instead, language (M (PI) = 3.19, M (AI) = 3.75) has a cluster center in the "adherents" area. This reveals that when it comes to language difference, "adherents" preferred to keep a closed mind instead of learning about it.

(5) *Neutralists*: Among "neutralists" education method is the most neutral cultural difference (M (PI) = 2.88, M (AI) = 2.94, N = 66).

- **Cluster size**

Like for the results of genre-related cultural difference, "(extreme) learners" and "neutralists" are the dominating viewer types, wheareas "escapees", adherents", and "adjusters" are subordnated.

*(1) Domination of "(extreme) learners" and "neutralists"*

For language, "extreme learners" (N = 92) and "neutralists" (N = 156) are dominate with 55.6% of the total sample size. Values comprise 93 "learners" and 144 "neutralists", which is 52.7% of the total sample size. The results indicate that one third of participants (N = 156 and 144) maintain a neutral attitude in regard to language and value difference. They do not pay much attention to it and do not care about it. One fifth of the participants (N = 92 and 93) keep a very open mind to learning about it.

=> H9, which assumed that the attitudes towards language difference are primarily those of "escapees" and "adherents", was not supported.

For way of life, "learners" (N = 102) and "extreme learners" (N = 160) dominate. Both groups account for 58.9% of the total sample size. Education method has the same amout of "learners" (N = 102) and "extreme learners" (N = 160). These results correspond to the interviewees statements. They preferred the foreign way of life and education methods presented in foreign programs.

=> H10a, which assumed that for the cultural differences of way of life and education methods viewers are primarily "learners", was strongly supported.

*Table 27: Cluster Description of Genre-Unrelated Cultural Difference*

| | | Language | | Value | | Way of life | | Education meth. | |
|---|---|---|---|---|---|---|---|---|---|
| | | | N | | N | | N | | N |
| Escapee | PI | 4.59 | 86 | 3.95 | 84 | 4.22 | 53 | 4.58 | 68 |
| | AI | 4.09 | | 3.18 | | 3.95 | | 3.55 | |
| Learner | PI | | | 3.49 | 67 | 4.55 | 102 | 3.83 | 102 |
| | AI | | | 1.59 | | 2.53 | | 2.28 | |
| Extreme Learner | PI | 4.03 | 92 | 4.33 | 93 | 4.33 | 160 | 4.70 | 160 |
| | AI | 1.66 | | 1.85 | | 1.32 | | 1.35 | |
| Adherent | PI | 3.19 | 34 | | | | | | |
| | AI | 3.75 | | | | | | | |
| Adjuster | PI | 2.20 | 78 | 2.28 | 62 | 1.55 | 31 | 2.14 | 49 |
| | AI | 1.83 | | 1.82 | | 1.47 | | 1.48 | |
| Neutralist | PI | 3.37 | 156 | 3.16 | 144 | 2.98 | 99 | 2.88 | 66 |
| | AI | 2.79 | | 2.71 | | 2.60 | | 2.94 | |
| Total | N | | 446 | | 450 | | 445 | | 445 |

Note: Education meth. = Education method.
Source: Own illustration.

*(2) Subordination of "escapees", "adherents", and "adjusters"*

For language, "adherents" are the smallest group with only 34 participants. For the other three dimensions, "adjusters" represent the smallest group (N = 62, 31, 49).

=> H7, which assumed that the viewer groups of “escapees”, “adherents”, and “adjusters” are subordinate to “learners”, was also supported by genre-unrelated cultural differences.

# 16 Results III: Influential Factors

In this chapter the research results about relationships between the perception/ acceptance indices and various influential factors are presented. Firstly, a hierarchical regression model is employed in order to analyze the impacts of four blocks of influential factors on the perception and acceptance indices. Then, several t-tests examine the impact of each influential factor separately. The results are presented below.

## 16.1 All Influential Factors

To explore the influential factor effects on the perception and acceptance of cultural differences, a hierarchical regression analysis was performed using 11 control variables for four steps. Each variable might have unique effects on perception and acceptance in addition to their joint predictive power. Table 28 shows the results of the hierarchical regression analysis. It examined the extent to which various influential factors contribute to a model of perception/acceptance degree.

All socio-demographic control variables (gender, age, education, income) were included in the first block, followed by the foreign culture exposure variables (knowledge, viewing) in the second, then the cultural proximity terms in the third (America, Europe, Asia, others), and finally the regional difference (city of residence) in the fourth block. The regression model used "listwise deletion" (N = 415) to deal with missing data. (See Table 28)

- **Perception**

Education ($\beta = .14$, $p < .01$), America ($\beta = -.30$, $p < .001$), and Asia ($\beta = .19$, $p < .001$) were three most influential factors in predicting participants' perceptions of cultural difference ($R = .37$, adjusted $R^2 = .117$, $p < .001$). Other influential factors did not predict participants' perceptions at a statistically significant level. (See Table 28)

The results of the hierarchical regression analysis indicate that multiple correlation coefficients were significantly different to zero at the end of each step – Step 1 ($R = .22$, adjusted $R^2 = .040$, $p < .001$), Step 2 ($R = .23$, adjusted $R^2 = .037$, $p < .01$), Step 3 ($R = .37$, adjusted $R^2 = .116$, $p < .001$), and Step 4 ($R = .37$, adjusted $R^2 = .117$, $p < .001$).

*Stage 1*: Gender, age, education, and income jointly accounted for 4.0% of the variance. Education ($\beta = .14$, $p < .01$) was able to predict positive attitudes towards the perception of cultural difference. This indicates that the higher the participants' education level, the greater the cultural difference they perceive.

*Step 2*: The independent variables jointly accounted for 3.7% of the variance in perception. After the entry of knowledge and viewing variables at Stage 2, none had significant effects on perception.

*Step 3*: The addition of cultural proximity variables to the equation resulted in a significant increment in adjusted $R^2$. The four proximity variables accounted for an extra 7.9% of the variance. They had significantly predicted the perception of cultural difference.

The preference for foreign television programs from America ($\beta = -.30$, $p < .001$) was strongly able to predict negative attitudes towards the perception of cultural difference. This indicates that the more the participants like watching programs from America, the greater the cultural difference they perceive.

In contrast, the preference for foreign programs from Asia ($\beta = .19$, $p < .001$) was able to predict positive attitudes towards the perception of cultural difference. This indicates that the more the participants like watching programs from Asia, the smaller the cultural difference they perceive.

*Step 4*: The addition of the city variable only accounted for an extra .1% of the variance. Thus, city of residence is not a significant variable to predict the perception index.

In total, 11 influential factors at four steps accounted for 11.7% of variance ($R = .37$, adjusted $R^2 = .117$, $p < .001$). America was the strongest predictor, with Asia second, and education third. The results indicate that participants with a higher education level, who do not like American programs but like Asian ones, would be more likely to perceive cultural difference as being large.

- **Acceptance**

Knowledge ($\beta = .17$, $p < .01$), America ($\beta = .30$, $p < .001$), and Europe ($\beta = .15$, $p < .05$) were three most influential factors in predicting participants' acceptance of cultural difference ($R = .51$, adjusted $R^2 = .243$, $p < .001$). Other influential factors did not predict participants' acceptance of cultural difference to a significant level. (See Table 28)

The results of the hierarchical regression analysis indicate that multiple correlation coefficients were also significantly different to zero at the end of each step, Step 1 ($R = .20$, adjusted $R^2 = .032$, $p < .01$), Step 2 ($R = .37$, adjusted $R^2 = .127$, $p < .001$), Step 3 ($R = .51$, adjusted $R^2 = .244$, $p < .001$), and Step 4 ($R = .51$, adjusted $R^2 = .243$, $p < .001$).

*Stage 1*: Gender, age, education, and income jointly accounted for 3.2% of the variance in acceptance. None of the demographic characteristics are significant for predicting participants' acceptance.

*Step 2*: Knowledge and viewing jointly accounted for 12.7% of variance. The addition of both variables to the equation resulted in a significant increment in adjusted $R^2$. They accounted for an extra 9.5% of the variance in acceptance.

Knowledge ($\beta = .17$, $p < .01$) was able to predict positive attitudes towards the acceptance of cultural difference. This indicates that the more viewers know about foreign culture, the more cultural difference they can accept. Viewing does not have a significant impact on the acceptance index.

*Step 3*: The addition of cultural proximity variables to the equation resulted in a more significant increment in $R^2$. Four proximity variables accounted for an extra 11.7% of the variance in acceptance. They significantly predicted the acceptance of cultural difference.

The preference for foreign television programs from America ($\beta = .30$, $p < .001$) was strongly able to predict positive attitudes towards the acceptance of cultural difference. This indicates that the more the participants like watching programs from America, the more cultural difference they can accept.

Similarly, the preference for programs from Europe ($\beta = .15$, $p < .001$) was also able to predict positive attitudes towards acceptance. The more the participants like watching programs from Europe, the more cultural difference they can accept.

In total, 11 influential factors in four steps accounted for 24.3% of the variance ($R = .51$, adjusted $R^2 = .243$, $p < .001$). America was the strongest predictor, knowledge the second, and Asia third. The results indicated that if participants like America or European programs, or possess a lot of knowledge about foreign cultures, they would be more likely to accept cultural difference than those who do not.

- **Cultural proximity**

On the whole, the beta coefficients indicate that cultural proximity variables were the most influential predictors in the regression equation to account for variance in both the perception and acceptance indices. Among the four preference variables, "America" and "Europe" could be considered as *culturally distant countries*, while "Asia" could be considered as *culturally proximate countries*.

The perception index is significantly related to the preference variables "America" and "Asia", while the acceptance index is significantly related to "America" and "Europe": (See Table 28)

*Table 28: Hierarchical Regression Explaining Perception/Acceptance Index*

| | Perception | | | Acceptance | | |
|---|---|---|---|---|---|---|
| Variable | B | β | T | B | β | T |
| *Block 1* | | | | | | |
| Gender | .13 | .10 | 1.96 | -.03 | -.03 | -.56 |
| Age | .06 | .04 | .70 | .08 | .05 | 1.00 |
| Education | .17 | .14 | 2.81** | .00 | .00 | .00 |
| Income | -.02 | -.02 | -.33 | -.02 | -.03 | -.54 |
| *Block 2* | | | | | | |
| Knowledge | -.02 | -.02 | -.38 | .15 | .17 | 3.15** |
| Viewing | -.03 | -.04 | -.58 | -.03 | -.04 | -.60 |
| *Block 3* | | | | | | |
| America | -.27 | -.30 | -4.02*** | .27 | .30 | 4.34*** |
| Europe | .06 | .06 | .76 | .14 | .15 | 2.00* |
| Asia | .16 | .19 | 3.65*** | -.01 | -.01 | -.23 |
| Other | .15 | .13 | 1.93 | .01 | .01 | .13 |
| *Block 4* | | | | | | |
| City | .02 | .06 | 1.14 | .02 | .04 | .78 |
| | | | | | | |
| Model | R | $\Delta R^2$ | F | R | $\Delta R^2$ | F |
| Block 1 | .22 | .040 | 5.33*** | .20 | .032 | 4.44** |
| Block 2 | .23 | .037 | 3.63** | .37 | .127 | 11.05*** |
| Block 3 | .37 | .116 | 6.44*** | .51 | .244 | 14.36*** |
| Total | .37 | .117 | 5.98*** | .51 | .243 | 13.09*** |

Note: N = 415. Listwise deletion. * $p < .05$. ** $p < .01$. *** $p < .001$.
Source: Own illustration.

(1) For *perception* the beta coefficient of America ($\beta = -.30$, $p < .001$) is negative, while for Asia ($\beta = .19$, $p < .001$) it is positive. In other words, participants who like programs from culturally proximate countries (Asia) and/or do not like

programs from culturally distant countries (America and Europe) perceive less cultural difference.

(2) For *acceptance* both, the beta coefficients of America ($\beta = .30$, $p < .001$) and Europe ($\beta = .15$, $p < .05$) were positive. This indicates that cultural proximity had a high impact on the perception and acceptance of cultural difference. This means that participants who like programs from culturally distant countries (America and Europe) could accept more cultural difference than who do not.

It is important to point out that the impact of the America variable on the perception index ($\beta = -.30$, $p < .001$) and the acceptance index ($\beta = .30$, $p < .001$) is the opposite. The both beta coefficients indicate that the more the participants like watching America television programs, the larger the cultural difference they can perceive, but the more cultural difference they can accept.

In sum, the overall predictive value of this regression model was relatively low, explaining just 11.7% of the variance for perception and 24.3% of the variance for acceptance. The perception and acceptance indices are influenced by several influential factors simultaneously.

The analysis of all influential factors as a whole in a model aims to provide an overview of the extent to which various influential factors contribute to this model. On its own the hierarchical regression model is not enough to investigate the relationships between influential factors and particular cultural dimensions. So, in the following, each influential factor is investigated separately.

## 16.2 Particular Influential Factors

The hypotheses about influential factors suggested that there was a significant attitudinal difference between demographic groups. In this part, t-tests and regression analysis were used to illustrate a possible diversity of participants' perception and acceptance of cultural difference; that is, in 10 dimension sets and 42 particular dimensions. T-tests were conducted in order to examine the impact of "gender", "age", and "city of residence" on perceptions and acceptance of cultural difference, while regression analysis was conducted to examine impact of other influential factors including "education", "income", "knowledge", "viewing", and "cultural proximity".

Table 29 presents the results of a correlation analysis of all influential factors:

*(1) Demographics:*

- *Age* was positively correlated to income ($\beta = .37$, $p < .01$). This meant the older the participants were, the higher their income.

- *Education* is negatively correlated to knowledge ($\beta = -.36$, $p < .01$) and preference for American programs ($\beta = -.30$, $p < .01$). This indicates that participants with higher education level know more about foreign culture and like American programs more than those with lower education.

- *Income* is negatively correlated to the ranking of city ($\beta$ = -.32, p < .01). This means the participants in more-open cities have higher income than those in more-conservative cities. (See Table 29)

*Table 29: Correlation Matrix for Influential Factors*

| | 2 | 3 | 4 | 5 | 6 | 7 | 8 | 9 | 10 | 11 |
|---|---|---|---|---|---|---|---|---|---|---|
| 1. Gender | -.11* | -.07 | -.14** | .14** | .30** | .16** | .13** | -.20** | .00 | -.02 |
| 2. Age | ------- | .04 | .37** | -.03 | -.11* | -.11* | -.09 | -.13* | -.15** | -.27** |
| 3. Education | ------- | ------- | .12* | -.36** | -.20** | -.30** | -.28** | .06 | -.07 | -.09 |
| 4. Income | ------- | ------- | ------- | -.16** | -.15** | -.11* | -.09 | .01 | -.10 | -.32** |
| 5. Knowledge | ------- | ------- | ------- | ------- | .49** | .44** | .40** | .13** | .32** | .11* |
| 6. Viewing | ------- | ------- | ------- | ------- | ------- | .49** | .51** | .21** | .45** | .17** |
| 7. America | ------- | ------- | ------- | ------- | ------- | ------- | .76** | .10 | .48** | .15** |
| 8. Europe | ------- | ------- | ------- | ------- | ------- | ------- | ------- | .15** | .64** | .13** |
| 9. Asia | ------- | ------- | ------- | ------- | ------- | ------- | ------- | ------- | .36** | .04 |
| 10. Other | ------- | ------- | ------- | ------- | ------- | ------- | ------- | ------- | ------- | .13** |
| 11. City | ------- | ------- | ------- | ------- | ------- | ------- | ------- | ------- | ------- | ------- |

Note: Listwise deletion. * p < .05. ** p < .01. N = 415.
Source: Own illustration.

*(2) Cultural exposure*

- *Knowledge* was positively correlated to viewing ($\beta$ = .49, p < .01) and to each cultural proximity variable. Among them, the correlation coefficients of America ($\beta$ = .44, p < .01), Europe ($\beta$ = .40, p < .01), and other non-Asian foreign countries ($\beta$ = .32, p < .01) was clearly higher than Asia ($\beta$ = .13, p < .01). This demonstrated that viewers who watch more foreign television programs, or like programs from America, Europe, or other non-Asian foreign countries, know more about foreign culture. (See Table 29)

- *Viewing* was also strongly positively correlated to America ($\beta$ = .49, p < .01), Europe ($\beta$ = .51, p < .01), and other non-Asian foreign countries ($\beta$ = .45, p < .01). These correlation coefficients were clearly higher than Asia ($\beta$ = .21, p < .01).

*(3) Cultural proximity*

- America was very strongly correlated to Europe ($\beta = .76, p < .01$) and other non-Asian countries ($\beta = .48, p < .01$). Europe was also very strongly correlated to other non-Asian countries ($\beta = .64, p < .01$). The positive correlation between America, Europe, and other non-Asian countries indicates that the participants who like to watch American programs usually also like to watch foreign programs from Europe and other non-Asian countries. (See Table 29)

- In contrast, the correlation between America and Asia ($\beta = .10$), Europe and Asia ($\beta = .15, p < .01$) was low. This indicated that participants who like to watch American programs usually do not like watching foreign programs from Asian countries.

## 16.2.1 Demographics: Gender, Age, Education, Income

- **Gender**

H11a and H11b suggested that the perception and acceptance of cultural difference displayed a significant difference between male and female viewers. Men *perceived* more cultural differences and *accepted* more cultural differences than women.

The results of the t-tests for the 10 cultural dimension sets and the 42 particular cultural dimensions demonstrate that generally there were no obvious differences between men and women. The difference only existed in a few aspects. "Pairwise deletion" was used to deal with missing data. (See Table 30)

(1) Comparing *10 cultural dimension sets*, only the acceptance of sports programs (total_sport) had a gender difference. (See Table 30) The mean value of male participants was 2.12 ($SD = .83, N = 197$), while the mean value of female participants was 2.45 ($SD = .97, N = 229$). The difference between the two mean values was very significant ($T = -3.79, p < .001$). This result indicated that males accept more cultural differences in foreign sports programs than females. The gender difference for the other 9 dimension sets was not significant.

(2) Comparing *42 particular cultural dimensions*, only five dimensions had a significant gender difference for acceptance. For "appearance of character" in foreign movies and television series females ($M = 1.98, SD = .97$) accept more than males ($M = 2.19, SD = 1.16$). This gender difference was significant ($T = 2.01, p < .05$) and the variance of the samples was also very significant ($F = 12.93, p < .001$).

Three particular dimensions of sports programs also had significant gender difference: "moderation/commentary" ($T = -3.85, p < .001$), "commercialization" ($T = -2.40, p < .05$), and "sports types" ($T = -3.86, p < .001$). For all three dimensions each mean value for men (M = 2.10, M = 2.32, M = 1.91) was smaller than for women (M = 2.49, M = 2.61, M = 2.30). This meant that males accept more cultural differences in foreign sports programs than females.

By "explicit vs. implicit" in documentaries gender difference was significant (T = -2.32, p < .05) with a significant variance (F = 12.03, p < .001). Men (M = 2.04, SD = .88) accept this cultural difference more than women (M = 2.25, SD = 1.01).

On the whole, there is no significant difference between male and female participants. Therefore:

=> H11a and H11b were not supported. In addition, H11b was only supported by sports programs and five particular cultural dimensions – appearance of characters (movies & TV series), moderation/commentary, commercialization, sports types (sports programs), and explicit vs. implicit (documentaries).

*Table 30: Gender as an Influential Factor*

| **Cultural Difference** | **Gender** | **N** | **M** | **SD** | **T** | **F** |
|---|---|---|---|---|---|---|
| *Movie & TV series* | | | | | | |
| (A) Appearance of character | M | 197 | 2.19 | 1.16 | 2.01* | 12.93*** |
| | F | 236 | 1.98 | .97 | | |
| *Sports programs* | | | | | | |
| (A) Total_sport | M | 197 | 2.12 | .83 | -3.79*** | 2.13 |
| | F | 229 | 2.45 | .97 | | |
| (A) Moderation/commentary | M | 195 | 2.10 | .97 | -3.85*** | 5.09* |
| | F | 227 | 2.49 | 1.12 | | |
| (A) Commercialization | M | 188 | 2.32 | 1.14 | -2.40* | 1.23 |
| | F | 216 | 2.61 | 1.21 | | |
| (A) Sports types | M | 195 | 1.91 | .99 | -3.86*** | 7.73** |
| | F | 227 | 2.30 | 1.12 | | |
| *Documentary* | | | | | | |
| (A) Explicit vs. implicit | M | 194 | 2.04 | .88 | -2.32* | 12.03*** |
| | F | 234 | 2.25 | 1.01 | | |

Note: (P) = perception; (A) = Acceptance. M = male; F = female. Paarwise deletion.
* < p < .05; ** p < .01; *** p < .001.

- **Age**

In this study there are three age groups: 20-29 (375 people, 83.3%), 30-39 (56 people, 12.4%), and 40-49 (11 people, 2.4%). In the t-tests the age groups 30-39 and 40-49 were combined together. The reasons are, firstly, that this study aims primarily to observe the generation difference between the post-80s generation (younger than 30) and the pre-80s generation (older than 30). Thus, the difference between the age groups 30-39 and 40-49 is not important. Secondly, the age group 40-49 has only 11 people, which makes it too small for a comparison with the other two age groups. After combining the groups, there were only two age groups remaining: 20-29 (375 people, 83.3%) and 30-49 (67 people, 14.8%). "Pairwise deletion" was used to deal with missing data.

H12a assumed that old viewers *perceive* more cultural difference than young viewers. H12b assumed that old viewers *accept* more cultural difference than young viewers.

Table 31 demonstrates the results of the t-tests. The results indicate that generally there was no difference between both age groups. Only two cultural dimensions have age difference: perception of "liberty/democracy/equality" ($T = 2.14$, $p < .01$) and acceptance of "patriotism" ($T = 2.29$, $p < .05$). Both belong to the dimension set "values".

*Table 31: Age as an Influential Factor*

| **Cultural Difference** | **Age** | **N** | **M** | **SD** | **T** | **F** |
|---|---|---|---|---|---|---|
| Value | | | | | | |
| (P) Liberty/democracy/equality | 20-29 | 370 | 3.44 | 1.23 | 2.14* | .002** |
| | 30-49 | 67 | 3.09 | 1.32 | | |
| (A) Patriotism | 20-29 | 366 | 2.31 | 1.09 | 2.29* | 5.872* |
| | 30-49 | 66 | 2.02 | .92 | | |

Note: (P) = perception; (A) = Acceptance. Paarwise deletion.
* < p < .05; ** p < .01; *** p < .001.
Source: Own illustration.

The age group 20-29 ($M = 3.44$, $SD = 1.23$) perceived cultural difference of "liberty/democracy/equality" more than age group 30-49 ($M = 3.09$, $SD = 1.32$), because in the living environment of this young generation they have more opportunities to learn about liberty, democracy, and equality in foreign countries through the media or visits abroad. However, young viewers of the age group 20-29 ($M = 2.31$, $SD = 1.09$) accepted less difference of "patriotism" than the older age group 30-49 (M = 2.02, SD = .92), because they have less of an awareness of patriotism than the older generation. (See Table 31)

=> H12a and H12b were not supported.

- **Education**

To examine the relationship between education and perception index as well as acceptance index, all participants were divided into three groups: primary/middle school (96 people, 21.8%), bachelor degree (309 people, 70.1%), and master's or over including doctorate (36 people, 8.1%).

H13a assumed that viewers with a high education level *perceive* more cultural difference than those with a low education level. H13b assumed that viewers with a high education level *accept* more cultural difference than those with a low education level.

Table 32 presents the results of the linear regression analysis for 10 dimension sets. In this table, "total" means a summary dimension whose perception index is the mean score of the perception indices of all 10 dimension sets and whose acceptance is the mean score of the acceptance indices of all 10 dimension sets. The regression model used "pairwise deletion" to deal with missing data.

(1) *Total value*: The results indicate that "education" was able to predict the "total" value of perception ($\beta = .22$, $p < .001$) and "total" value of acceptance ($\beta = -.19$, $p < .001$) at a statistically significant level. The overall predictive value of the regression model was relatively low, explaining only 4% of the variance for perception (adj. $R^2 = .04$) and 3% of the variance for acceptance (adj. $R^2 = .03$). The positive beta coefficient of perception ($\beta = .22$) indicates that viewers with a high education level accept more cultural difference than those with a low education level. The negative beta coefficient of acceptance ($\beta = -.19$) indicates that viewers with a high education level accept more cultural difference than those with a low education level. (See Table 32)

(2) *Perception*: Investigating the results of the regression analysis for 10 dimension sets, education as an influential factor was also able to predict the perception of almost all sets except "movies & TV series". The most influential impact of education is on "entertainment shows" ($\beta = .20$, $p < .001$) and "education method" ($\beta = .21$, $p < .001$). The variance varies between 1% and 4%. All of nine beta coefficients for perception were positive. This indicates that for these nine cultural dimension sets, the higher the participants' education, the more cultural difference they perceive. (See Table 32)

(3) *Acceptance*: Education was also able to predict the acceptance of almost all dimension sets except "language" and "values". The impact of education on "movies & TV series" ($\beta = -.24$, $p < .001$), "documentaries" ($\beta = -.23$, $p < .001$),

and “music programs” (β = -.24, p < .001) was the strongest. The variance varied between 1% and 6% of the samples. All of the eight beta coefficients for acceptance were negative. This indicated that for these eight dimension sets, the higher the participants’ education then the more cultural difference they can accept.

In sum, participants’ education level was able to predict perception and acceptance significantly and strongly, not only for the “total” summary dimension, but also for most of the dimension sets. Education level was positively correlated to perception index and negatively correlated to acceptance index.

=> Both H13a and H13b were supported for the total value. In addition, H13a was supported for nine dimension sets and H13b was supported for eight dimension sets.

*Table 32: Education as an Influential Factor*

| | Perception | | | | Acceptance | | | |
|---|---|---|---|---|---|---|---|---|
| | β | $\Delta R^2$ | F | N | β | $\Delta R^2$ | F | N |
| Movie & TV series | | | | | -.24 | .06 | 27.35*** | 436 |
| Sport | .14 | .02 | 8.83** | 433 | -.12 | .01 | 6.41* | 426 |
| Documentary | .12 | .01 | 6.43* | 436 | -.23 | .05 | 23.17*** | 431 |
| Entertainment | .20 | .04 | 17.27*** | 433 | -.11 | .01 | 4.83* | 433 |
| News | .13 | .01 | 6.72** | 427 | -.15 | .02 | 9.35** | 426 |
| Music | .14 | .02 | 8.62** | 432 | -.24 | .06 | 26.66*** | 427 |
| Language | .19 | .03 | 15.66*** | 440 | | | | |
| Value | .17 | .03 | 12.80*** | 441 | | | | |
| Way of life | .13 | .01 | 6.91** | 436 | -.16 | .02 | 10.82*** | 436 |
| Education | .21 | .04 | 20.15*** | 436 | -.11 | .01 | 4.95* | 436 |
| Total | .22 | .04 | 21.40*** | 441 | -.19 | .03 | 16.42*** | 441 |

Note: Paarwise deletion. * p < .05. ** p < .01. *** p < .001.
Source: Own illustration.

- **Income**

The participants comprised four income groups: ¥ 0-999 (105 people, 23.3%), ¥ 1000-4999 (264 people, 58.7%), and over ¥ 5000 (55 people, 12.2%).

H14a assumed that viewers with a high income *perceive* more cultural difference than those with a low income. H14b assumed that viewers with a high income *accept* more cultural difference than those with a low income.

Table 33 demonstrates the results of the linear regression analysis of the 10 dimension sets and a "total" summary dimension. The regression model used "pairwise deletion" to deal with missing data.

The results presented that income as a control factor was not able to predict perception and acceptance on the whole. It was only able to predict perception of "sports programs" ($\beta = -.11$, $p < .05$) and "documentaries" ($\beta = -.15$, $p < .01$). Income accounted for only 1% of variance in sports programs and 2% in documentaries. Both beta coefficients meant that the higher the participants' income is the more cultural difference in sports programs and documentaries they can accept.

=> H14a and H14b were not supported. In addition, two exceptions for H14b were sports programs and documentaries.

*Table 33: Income as an Influential Factor*

| | Perception | | | | Acceptance | | | |
|---|---|---|---|---|---|---|---|---|
| | $\beta$ | $\Delta R^2$ | F | N | $\beta$ | $\Delta R^2$ | F | N |
| Sport | | | | | -.11 | .01 | 4.87* | 410 |
| Documentary | | | | | -.15 | .02 | 9.55** | 414 |

Note: Paarwise deletion. * $p < .05$. ** $p < .01$. *** $p < .001$.
Source: Own illustration.

- **Summary**

Based on the results of the t-tests and regression analysis, among the four demographic characters "education" remained the most important in accounting for participants' perception and acceptance towards cultural difference – not only for the "total" values, but also for most particular dimension sets. Gender, age, and income influenced only a few dimension sets. They did not have a significant influence on the "total" value. These findings indicated that the degree of perception and acceptance depended primarily on which education level viewers had attained.

### 16.2.2 Knowledge about Foreign Culture

Knowledge about foreign culture means insights into foreign society. H15a assumed that viewers who know a lot about foreign culture *perceive* more cultural difference than those who know little. H15b assumed that viewers who know a great deal about foreign culture *accept* more cultural difference than those who know only a little about it.

Among the five participant groups regarding knowledge about foreign culture – "very much", "much", "normal", "little", and "very little" – the "normal" group (N = 263, 58.4%) was the largest. The other two groups "much" (N = 59, 13.1%) and "little" (N = 90, 20.0%) accounted for 33.1% of participants. The smallest groups were "very much" (N = 4, .9%) and "very little" (N = 18, 4.0%).

The results of the t-test indicated that these five groups were significantly different to each other (T = 89.84, $p < .000$). The mean score of all five groups was M = 3.14 (SD = .73), which was slightly larger than the middle value of scale 3. Table 34 presents the results of the linear regression analysis. The regression model used "pairwise deletion" to deal with missing data.

(1) *Total value*: Knowledge was able to significantly predict the "total" *perception* index, but not very strongly ($\beta = -.10$, adj. $R^2 = .01$, $p < .05$). Knowledge accounted for only 1% of the variance in "total" perception. This negative beta coefficient indicated that viewers who know a lot about foreign culture perceive more cultural difference than others who know little. (See Table 34)

Knowledge was also able to predict the "total" *acceptance* index not only significantly, but also strongly ($\beta = .34$, adj. $R^2 = .11$, $p < .001$). The positive beta coefficient indicated that viewers who know a great deal about foreign culture accept more cultural difference than others who know little.

The overall predictive value of the regression model for acceptance (adj. $R^2 = .11$) clearly explained more variance than perception (adj. $R^2 = .01$). Thus, H15a and H15b were supported for total value.

(2) *Particular dimension sets*: The variable knowledge was only able to predict the *perception* index of "sports programs" ($\beta = -.13$, adj. $R^2 = .01$, $p < .01$) and "entertainment shows" ($\beta = -.11$, adj. $R^2 = .01$, $p < .05$) significantly, but not strongly. Both beta coefficients were negative. Thus, H15a was only supported for sports programs and entertainment shows. (See Table 34)

In contrast, knowledge was able to predict the acceptance indices of all 10 particular cultural dimensions. Its impact on "documentaries" ($\beta = .35$, adj. $R^2 = .12$, $p < .001$), "music programs" ($\beta = .31$, adj. $R^2 = .09$, $p < .001$), and "way of life" ($\beta = .30$, adj. $R^2 = .09$, $p < .001$) was stronger than for the other dimensions. The beta coefficients of all 10 dimensions were positive. The variance varied between .01 and .12 for each dimension. Thus, H15b was supported for all 10 cultural dimensions.

=> H15a and H15b were supported for both total value of perception and acceptance. In addition, H15a was only supported for two particular dimensions

– sports programs and entertainment shows. H15b was supported for all ten dimensions.

On the whole, a high level of knowledge of foreign culture is combined with a high degree of perception and a high degree of acceptance. It is a very important control factor, particularly for predicting acceptance index.

*Table 34: Knowledge as an Influential Factor*

| | Perception | | | | Acceptance | | | |
|---|---|---|---|---|---|---|---|---|
| | β | $\Delta R^2$ | F | N | β | $\Delta R^2$ | F | N |
| Movie & TV series | | | | | .27 | .07 | 32.19*** | 429 |
| Sport | -.13 | .01 | 6.98** | 426 | .25 | .06 | 27.72*** | 419 |
| Documentary | | | | | .35 | .12 | 57.77*** | 424 |
| Entertainment | -.11 | .01 | 5.23* | 426 | .19 | .04 | 16.32*** | 426 |
| News | | | | | .21 | .04 | 18.41*** | 419 |
| Music | | | | | .31 | .09 | 43.39*** | 421 |
| Language | | | | | .10 | .01 | 4.07* | 431 |
| Value | | | | | .25 | .06 | 27.69*** | 434 |
| Way of life | | | | | .30 | .09 | 43.44*** | 429 |
| Education | | | | | .22 | .05 | 22.60*** | 429 |
| Total | -.10 | .01 | 4.53* | 434 | .34 | .11 | 55.99*** | 434 |

Note: Paarwise deletion. * $p < .05$. ** $p < .01$. *** $p < .001$.
Source: Own illustration.

## 16.2.3 Viewing Foreign Television Programs

In this part both hypotheses suggested that there was significant attitudinal difference between viewers who watch foreign television programs and those who did not. H16a assumed that viewers who watch a lot of foreign programs *perceive* more cultural difference than those who watch few. H16b assumed that viewers who watch a lot of foreign programs *accept* more cultural difference than those who watch only a few.

To examine participants' viewing foreign programs more precisely, in the survey they were asked to evaluate how many foreign programs they watched for each television genre. Viewing might vary significantly between television genres, for example, viewing news might not be equal to viewing movies.

Therefore, the viewing index must be differentiated for genre-related and genre-unrelated dimensions.

- In linear regression models of six *genre-related* dimensions, the viewing index of each genre was used to examine the relationships between viewing of each genre and the perception/acceptance of each genre.
- In contrast, in the regression models of four *genre-unrelated* dimensions, the mean score of the viewing index of all six *genre-related* dimensions was used to examine the relationships between "total" viewing and the perception/acceptance of each genre-unrelated dimension.

Among the six television genres, foreign "movies & TV series" were viewed the most. Its mean score ($M = 2.44$, $SD = 1.03$) was between 2 and 3. This meant that generally participants watched "a lot" of movies and TV series. The mean scores of the other five genres varied between 3.22 and 3.88. This indicates that participants watched these five genres "little". Viewing "sports programs", "documentaries", "entertainment programs" and "music programs" was almost at an equal level. Viewing "news" ($M = 3.88$, $SD = .94$) took place the least. On the whole, viewing of all six genres ($M = 3.26$, $SD = .69$) was relatively "little". The results of the t-test indicated that the t-values of all six genres and a total value of six genres are at a significant level ($p < .000$). (See Table 35)

*Table 35: Descriptive Data of Viewing*

| | M | SD | T | N |
|---|---|---|---|---|
| Movie & TV series | 2.44 | 1.03 | 50.11*** | 446 |
| Sport | 3.22 | 1.29 | 53.11*** | 450 |
| Documentary | 3.23 | 1.06 | 64.68*** | 446 |
| Entertainment | 3.48 | 1.00 | 73.17*** | 443 |
| News | 3.88 | .94 | 87.17*** | 442 |
| Music | 3.33 | 1.07 | 65.37*** | 438 |
| Total | 3.26 | .69 | 100.31*** | 450 |

Note: Paarwise deletion. * $p < .05$. ** $p < .01$. *** $p < .001$.
Source: Own illustration.

Table 36 presents the results of the linear regression analysis (pairwise deletion):

(1) *Total value*: Similar to "knowledge", "viewing" as an influential factor was also not significantly correlated to the perception index, but was strongly correlated to the acceptance index and able to significantly predict the ac-

ceptance index ($\beta$ = .25, adj. $R^2$ = .06, p < .001). The beta coefficient of acceptance was positive. This indicated that viewers who watch a lot of foreign programs accept more cultural differences than those who watch few. Viewing accounted for 6% of variance in acceptance. Thus, H16a was not supported for the total value, but H16b was supported for the total value.

*Table 36: Viewing as an Influential Factor*

| | Perception | | | | Acceptance | | | |
|---|---|---|---|---|---|---|---|---|
| | $\beta$ | $\Delta R^2$ | F | N | $\beta$ | $\Delta R^2$ | F | N |
| Movie & TV series | -.18 | .03 | 14.26*** | 445 | .24 | .05 | 26.10*** | 441 |
| Sport | -.10 | .01 | 3.97* | 442 | .30 | .09 | 42.58*** | 435 |
| Documentary | -.12 | .01 | 6.83** | 441 | .28 | .08 | 37.93*** | 436 |
| Entertainment | | | | | .25 | .06 | 29.50*** | 434 |
| News | | | | | | | | |
| Music | | | | | .22 | .05 | 22.13*** | 426 |
| Language | | | | | | | | |
| Value | | | | | .19 | .03 | 16.57*** | 450 |
| Way of life | | | | | .21 | .04 | 19.77*** | 445 |
| Education | | | | | .14 | .02 | 9.10** | 445 |
| Total | | | | | .25 | .06 | 29.78*** | 450 |

Note: Paarwise deletion. * p < .05. ** p < .01. *** p < .001.
Source: Own illustration.

(2) *Particular dimensions*: Investigating particular dimensions viewing was only able to predict the perception index of "movies & TV series" ($\beta$ = -.18, p < .001), "sports programs" ($\beta$ = -.10, p < .05), and "documentaries" ($\beta$ = -.12, p < .01). The adjusted $R^2$ is low and varies between .01 and .03. All three beta coefficients were negative. This meant that for these three genres, the more foreign programs viewers watch the more cultural difference they perceive.

In contrast, viewing was able to predict the acceptance index of almost all dimensions except "news" and "language". The cultural dimensions, which were correlated to viewing the strongest, were "sports programs" ($\beta$ = .30, adj. $R^2$ = .09, p < .001) and "documentaries" ($\beta$ = .28, adj. $R^2$ = .08, p < .001). The beta coefficients of all eight dimensions were positive. This meant that the more foreign programs viewers watch, the more cultural difference they accept.

=> Thus, H16a was not supported for the total value, but H16b was supported for the total value. In addition, H16a was supported for three particular dimensions and H16b was supported for eight cultural dimensions.

On the whole, viewing is not combined with perception degree, but a lot of viewing is combined with a high acceptance degree. Viewing of foreign programs is a very important control factor for predicting the acceptance index significantly and strongly.

## 16.2.4 Cultural Proximity

In this part, five hypotheses are examined to determine if participants' cultural preference is related to perception and acceptance index. To examine participants' cultural preference, in the survey they were asked to evaluate how much they like foreign television programs from certain countries for each television genre. Cultural preference might vary significantly across television genres, for example, cultural preference for television series might not be equal to sports programs. They have four options of country of origin: America, Europe, Asia, and other countries. Therefore, the cultural preference indices must be differentiated for different genres:

- *Genre-related dimensions*: In the regression models of the six genre-related dimensions, the cultural preference index of each television genre was used to examine relationship between the cultural preference and the perception/acceptance of each genre.

- *Genre-unrelated dimensions*: In the regression models of the four genre-unrelated dimensions, a "total" preference index was used to examine the relationship between "total" index and perception/acceptance of each genre-unrelated dimension. The "total" index is the mean score of the preference indices of all six genre-related dimensions.

Table 37 presents some descriptive data:

(1) *Total value*: Chinese participants *like* programs from "America" (M = 2.54, SD = .70) the most. They *like* "Asian" programs (M = 2.86, SD = .75) and those from "Europe" (M = 2.80, SD = .69) almost at an equal level. They *do not like* programs from "other" foreign countries (M = 3.16, SD = .54).

(2) *Particular genres*:

- Among programs from "Asian" countries, participants *like* to watch "movies & TV series", "music programs", and "entertainment shows" (2 < M < 3). They *do not like* to watch "sports programs", "documentaries", and "news" (3 < M < 4).

- Among programs from "America", the mean score of all six genres were between 2 and 3. This indicated that participants *like* all kinds of American programs. The most preferred genre was "movies & TV series" (M = 2.32, SD = 1.02).

- Like for America, participants also *like* all kinds of programs from European countries (2 < M < 3). The most preferred genre was "music programs" (M = 2.62, SD = .95).

- They *do not like* programs from "other" non-Asian countries (3 < M < 4).

=> H17, which assumed that viewers like television programs from culturally proximate countries (Asia) than culturally distant countries (America, Europe, and other non-Asian countries), was not supported by all genres except "entertainment shows".

These results indicate that participants like programs from culturally distant America and European countries more than Asian countries. Only regarding entertainment shows, which need greater cultural proximity, did viewer like only the ones from Asia. Generally, how much viewers like foreign programs is primarily related to how successful the programs are on the global market (e.g., American movies), and not cultural proximity between importing and exporting countries.

*Table 37: Descriptive Data of Cultural Proximity*

| | Asian | | America | | Europe | | Others | |
|---|---|---|---|---|---|---|---|---|
| | M | SD | M | SD | M | SD | M | SD |
| Movie & TV series | 2.48 | 1.07 | **2.32** | **1.02** | 2.83 | .94 | 3.20 | .74 |
| Sport | 3.23 | .98 | **2.51** | **.99** | 2.75 | .99 | 3.23 | .72 |
| Documentary | 3.07 | 1.02 | **2.49** | **.99** | 2.74 | .97 | 3.10 | .81 |
| Entertainment | **2.67** | **1.03** | 2.73 | .91 | 2.98 | .82 | 3.19 | .73 |
| News | 3.09 | .94 | **2.77** | **.89** | 2.93 | .89 | 3.18 | .78 |
| Music | 2.57 | .99 | **2.43** | **.91** | 2.62 | .95 | 3.04 | .79 |
| Total | **2.86** | **.75** | **2.54** | **.70** | **2.80** | **.69** | 3.16 | .54 |

Source: Own illustration.

- **Culturally proximate countries – Asia**

H18a assumed that viewers who like watching TV programs from culturally proximate countries (Asia) *perceive* less cultural difference than those who do not. H18b assumed that viewers who like watching TV programs from culturally proximate countries (Asia) *accept* less cultural difference than those who do not.

(1) *Total value*: The results of the regression analysis in Table 38 indicate that the preference for television programs from Asian countries (except China) was a significant predictor for the perception index at a significant level ($\beta$ = .20, adj. $R^2$ = .04, $p < .001$). This positive beta coefficient indicated that viewers who like watching programs from culturally proximate countries (Asia) perceive less cultural difference than those who do not. H18a was supported for the total value.

The results of the regression analysis also showed that the preference for Asian programs was not a significant predictor for the acceptance index. H18b was not supported for total value.

(2) *Particular dimensions*: Asia was significantly correlated to the perception indices of seven particular dimensions. All of these seven beta coefficients were positive. The largest were "language" ($\beta$ = .20, $p < .001$) and "education method" ($\beta$ = .18, $p < .001$). This finding indicated that the more viewers like Asian television programs, the smaller the cultural difference of language and education method they perceive. H18a was supported for seven cultural dimensions.

Asia was only significantly correlated to the acceptance index of the three dimensions "entertainment shows" ($\beta$ = .23, $p < .001$), "language" ($\beta$ = .13, $p < .01$), and "values" ($\beta$ = .12, $p < .05$). Their beta coefficients were positive. This indicated that viewers who like watching Asian programs accept more cultural difference in entertainment shows, language, and values than those who do not like watching them. This finding is runs counter to H18b. Thus, H18b was not supported for all particular dimensions.

=> H18a was supported for the total value. H18b was not supported for the total value. In addition, H18a was supported for seven particular dimensions and H18b was not supported for all dimensions.

In sum, preference for Asian television programs (except China) is a significant predictor for the perception index, but not significant for the acceptance index.

- **Culturally distant countries – America, Europe, and other countries**

H19a assumed that viewers who like watching TV programs from culturally distant foreign countries (America, Europe, and other non-Asian countries) *perceive* more cultural difference than those who do not. H19b assumed that viewers who like watching programs from culturally distant countries *accept* more cultural difference than those who do not.

*(1) Total value*

*Perception*: America (β = -.20, adj. $R^2$ = .04, p < .001) and Europe (β = -.10, adj. $R^2$ = .01, p < .05) were able to predict perception index. The predictor of America was stronger than Europe. America could explain 4% of the variance, and Europe only 1%. Different to Asia, both beta coefficients for America and Europe were negative. This finding indicated that the viewers who like watching programs from culturally distant foreign countries (America, Europe) perceive more cultural difference than those who do not. "Other" non-Asian countries could not predict perception index. (See Table 38) Thus, H19a was supported for the total value of America and Europe.

*Acceptance*: America (β = .46, adj. $R^2$ = .21, p < .001), Europe (β = .43, adj. $R^2$ = .18, p < .05), and other countries (β = .27, adj. $R^2$ = .07, p < .001) were very significant predictors for the acceptance index, which could predict acceptance much more strongly than perception. America accounted for 21% of variance, Europe 18%, and other countries only 7%. Three beta coefficients (β = .46, β = .43, and β = .27) were large and positive. This finding indicated that viewers who like programs from culturally distant foreign countries (America, Europe, others) accept more cultural difference than those who do not. (See Table 38) H19b was strongly supported for the total value.

*(2) Particular dimensions*

*Perception*: Examining the relationship between cultural preference and the perception index of particular cultural dimensions, the results demonstrate that: (See Table 38)

- "America" was significantly and negatively correlated to the perception index of seven cultural dimensions. There was no large difference between seven beta coefficients. The largest was "documentaries" (β = -.17, p < .001). This indicated that for these dimensions, viewers who like American programs perceive *more* difference than those who do not.

- "Europe" was significantly and negatively correlated to "sports programs" (β = -.11, p < .05) and "music programs" (β = -.10, p < .05). Neither beta coefficient was very strong. This meant that viewers who like European programs perceive *more* cultural difference than those who do not.

- "Other" countries were significantly, but positively correlated to "news" (β = .11, p < .05), "language" (β = .10, p < .05) and "music programs" (β = .11, p < .05). This finding indicated that viewers who like programs from other countries perceive *less* cultural difference than those who do not.

Therefore, H19a was supported for seven dimensions for America and two dimensions for Europe. H19a was not supported for other countries.

*Acceptance*: Examining the relationship between cultural preference and the acceptance index of particular cultural dimensions, the results demonstrate that

America, Europe, and other countries were able to predict the acceptance index of all ten particular dimensions. (See Table 38)

- Ten beta coefficients for America were positive. The largest were "documentaries" ($\beta = .40$, adj. $R^2 = .16$, $p < .001$), "music programs" ($\beta = .39$, adj. $R^2 = .15$, $p < .001$), and "movies & TV series" ($\beta = .38$, adj. $R^2 = .14$, $p < .001$).
- Ten beta coefficients for Europe and other countries were also positive. The largest one for Europe was "music programs" ($\beta = .42$, adj. $R^2 = .18$, $p < .001$). The largest ones for other countries were "music programs" ($\beta = .27$, adj. $R^2 = .07$, $p < .001$) and "values" ($\beta = .27$, adj. $R^2 = .07$, $p < .001$).

On the whole, the positive coefficients indicated that viewers who like programs from culturally distant foreign countries (America, Europe, and other non-Asian countries) accept more cultural difference than those who do not. Thus, H19b was strongly supported for all ten particular dimensions.

=> H19a was supported for the total value of America and Europe, but not supported for other non-Asian countries. H19b was supported for the total value. In addition, H19a was supported for seven dimensions for America and two dimensions for Europe. H19b was strongly supported for all ten dimensions.

In sum, preference for television programs from culturally distant America, Europe, and other non-Asian countries is only able to predict the perception index of a few cultural dimensions, whereas it is able to strongly predict the acceptance index of all cultural dimensions. It is important to point out that "music programs" is the television genre whose acceptance index could be predicted by America, Europe, and other countries the strongest.

*Table 38: Cultural Proximity as an Influential Factor*

| | Asia | | America | | Europe | | Others | |
|---|---|---|---|---|---|---|---|---|
| | β | $\Delta R^2$ | β | $\Delta R^2$ | β | $\Delta R^2$ | β | $\Delta R^2$ |
| *(1) Perception* | | | | | | | | |
| Movie & TV s. | | | -.12** | .01 | | | | |
| Sport | .16*** | .02 | | | -.11* | .01 | | |
| Documentary | .12* | .01 | -.17*** | .03 | | | | |
| Entertainment | | | | | | | | |
| News | .11* | .01 | | | | | .11* | .01 |
| Music | | | -.16*** | .02 | -.10* | .01 | | |
| Language | .20*** | .04 | -.16*** | .02 | | | .10* | .01 |

| | | | | | | | | |
|---|---|---|---|---|---|---|---|---|
| Value | .11* | .01 | -.14** | .02 | | | | |
| Way of life | .11* | .01 | -.12** | .01 | | | | |
| Education | .18*** | .03 | -.14** | .02 | | | .11* | .01 |
| Total | .20*** | .04 | -.20*** | .04 | -.10* | .01 | | |
| *(2) Acceptance* | | | | | | | | |
| Movie & TV s. | | | .38*** | .14 | .30*** | .09 | .13** | .01 |
| Sport | | | .35*** | .12 | .33*** | .11 | .20*** | .04 |
| Documentary | | | .40*** | .16 | .34*** | .11 | .21*** | .04 |
| Entertainment | .23*** | .05 | .34*** | .11 | .30*** | .09 | .24*** | .06 |
| News | | | .27*** | .07 | .26*** | .07 | .23*** | .05 |
| Music | | | .39*** | .15 | .42*** | .18 | .27*** | .07 |
| Language | .13** | .01 | .12** | .01 | .15** | .02 | .13** | .02 |
| Value | .12* | .01 | .32*** | .10 | .32*** | .10 | .27*** | .07 |
| Way of life | | | .35*** | .12 | .34*** | .11 | .22*** | .05 |
| Education | | | .32*** | .10 | .27*** | .07 | .16*** | .02 |
| Total | | | .46*** | .21 | .43*** | .18 | .27*** | .07 |

Note: Paarwise deletion. * $p < .05$. ** $p < .01$. *** $p < .001$.
Perception: $419 \leq N$ (Asia) $\leq 450$, $434 \leq N$ (America) $\leq 450$, $428 \leq N$ (Europe) $\leq 450$, $412 \leq N$ (others) $\leq 448$.
Acceptance: $431 \leq N$ (Asia) $\leq 450$, $417 \leq N$ (America) $\leq 450$, $416 \leq N$ (Europe) $\leq 450$, $411 \leq N$ (others) $\leq 450$.
Source: Own illustration.

## 16.2.5 Regional Difference

H20a assumed that viewers from more-open cities perceive greater cultural difference than those from more-conservative cities. H20b assumed that viewers from more-open cities accept more cultural difference than those from more-conservative cities. In this study Beijing, Shanghai, and Guangzhou are regarded as more-open cities. Nanjing and Xi'An as two cultural cities with long histories are regarded as more-conservative cities.

Firstly, one-way ANOVA was used to examine whether five cities were significantly different to each other:

*(1) All five cities*

*Total value*: The mean values of the total perception index ($F = 2.58$, $p < .05$) of the five cities were different at a statistically significant level. The mean values of the total acceptance index ($F = 5.07$, $p < .001$) showed more of a difference to each other than the perception index. (See Table 39)

*Particular dimensions*: The perception index of "news programs" ($F = 2.91$, $p < .05$) and "language" ($F = 2.98$, $p < .05$) were significantly different in the five cities. The acceptance index of almost all ten dimensions was also significantly different in the five cities, except "sports programs" and "entertainment shows". (See Table 39)

*(2) Particular city*

*Perception*: Investigating the five cities separately, the perception index of Nanjing ($M = 3.73$) as a conservative city was the highest for almost all of the ten dimensions. The perception index of the other four cities did not display large differences between each other. This finding indicated that generally cultural difference in the minds of viewers from Nanjing was greater than for viewers from the other four cities. (See Table 39)

*Acceptance*: The acceptance index of Guangzhou ($M = 2.47$) was the highest and Beijing ($M = 2.04$) the lowest. This finding demonstrated that although Guangzhou was a more-open city, the viewers from this city had the lowest degree of acceptance. The viewers from Beijing, as a more-open city, could accept cultural difference the most.

Secondly, correlation analysis was used to examine how the five cities were different from each other; in other words, if viewers from more-open cities perceive and accept more cultural difference than those from more-conservative cities.

(1) *Total value*: Perception index was not significantly correlated with city ranking, whereas acceptance index was significantly correlated with city ranking ($\beta = .12$, $p < .01$). The positive beta coefficient indicated that viewers who live in more-open cities accept more cultural difference than those in more-conservative cities. (See Table 39)

(2) *Particular dimensions*: The correlation coefficient was not significant for the perception index of all ten dimensions. Only the acceptance index of "sports programs" ($\beta = .11$, $p < .05$), "documentaries" ($\beta = .18$, $p < .001$), "values" ($\beta = .12$, $p < .05$), and "way of life" ($\beta = .13$, $p < .01$) was significantly correlated to city ranking. These four beta coefficients were positive. This meant that for these four cultural differences there was a high level of difference between individuals' acceptance in each city.

*Table 39: Resident City as an Influential Factor*

| | BJ | SH | GZ | NJ | XA | F | β |
|---|---|---|---|---|---|---|---|
| *(1) Perception* | | | | | | | |
| Movie & TV series | 3.40 | 3.50 | 3.55 | 3.56 | 3.37 | | |
| Sports | 3.33 | 3.32 | 3.16 | 3.49 | 3.29 | | |
| Documentary | 3.48 | 3.22 | 3.25 | 3.44 | 3.29 | | |
| Entertainment | 3.56 | 3.66 | 3.56 | 3.93 | 3.62 | | |
| News | 3.34 | 3.21 | 3.25 | 3.59 | 3.18 | 2.91* | |
| Music | 3.50 | 3.42 | 3.32 | 3.74 | 3.63 | | |
| Language | 3.54 | 3.51 | 3.38 | 3.81 | 3.37 | 2.98* | |
| Value | 3.54 | 3.40 | 3.36 | 3.60 | 3.50 | | |
| Way of life | 3.89 | 3.75 | 3.79 | 4.07 | 3.83 | | |
| Education | 4.11 | 3.78 | 3.82 | 4.05 | 3.91 | | |
| Total | 3.57 | 3.48 | 3.44 | 3.73 | 3.50 | 2.58* | |
| *(2) Acceptance* | | | | | | | |
| Movie & TV series | 1.94 | 2.18 | 2.56 | 2.04 | 2.12 | 7.09*** | |
| Sport | 2.05 | 2.26 | 2.39 | 2.40 | 2.35 | | .11* |
| Documentary | 1.83 | 2.07 | 2.40 | 2.15 | 2.33 | 6.40*** | .18*** |
| Entertainment | 2.09 | 2.40 | 2.50 | 2.43 | 2.42 | | |
| News | 2.06 | 2.32 | 2.54 | 2.23 | 2.31 | 3.20* | |
| Music | 1.82 | 2.04 | 2.30 | 1.82 | 2.13 | 4.02** | |
| Language | 2.63 | 2.78 | 2.77 | 2.94 | 2.46 | 2.66* | |
| Value | 2.18 | 2.26 | 2.53 | 2.41 | 2.31 | 3.25* | .12* |
| Way of life | 1.88 | 2.21 | 2.37 | 2.05 | 2.36 | 3.98** | .13** |
| Education | 1.87 | 2.28 | 2.36 | 1.94 | 2.20 | 4.23** | |
| Total | 2.04 | 2.28 | 2.47 | 2.24 | 2.30 | 5.07*** | .12** |

Note: Listwise deletion. * < p < .05; ** p < .01; *** p < .001.
BJ = Beijing, SH = Shanghai, GZ = Guangzhou, NJ = Nanjing, XA = Xi'An.
N (perception) = 413; N (acceptance) = 399.
Source: Own illustration.

Therefore, on the one hand, although the perception indices of the five cities were significantly different to each other, they not significantly correlated to city ranking.

=> H20a was not supported.

On the other hand, the acceptance indices of the five cities were not only significantly different to each other, but also significantly correlated to city ranking. The more open a city is, the more cultural differences viewers could accept.

=> H20b was supported.

# 17 Summary of Part D3

A cultural difference should not be perceived and accepted uniformly. This part of the research concerning the perception and acceptance of cultural difference developed a quantitative method to examine three aspects of cultural difference. A total of 450 surveys were conducted in five sample cities. In the survey, 10 dimension sets (incl. 42 particular dimensions) were evaluated quantitatively. The results consist of three parts:

*(1) Degree of perception and acceptance*

Generally, although cultural differences were perceived to be large, viewers could mostly accept them.

Among *genre-related* dimensions "music programs" which contain little linguistic content was the most accepted television genre, whereas "news programs" in which speech was an important element was less acceptable.

Among *genre-unrelated* dimensions, "education method" and "way of life" were the two largest difference dimensions, but they were acceptable the most. This finding indicated that viewers looked forward to adopting an education method and way of life that they had seen in foreign programs. In contrast, "language" difference was the most unacceptable, especially "accuracy of translation". Thus, most viewers preferred to watch foreign programs in the original language more than dubbed.

Cultural level was not correlated to the perception and acceptance indices. This meant that differences at a deep level were not necessarily perceived to be larger and more unacceptable than those at surface level.

*(2) Stages of viewers' cultural adaptation*

A scatter diagram was used to demonstrate the five viewer types. Among them "(extreme) learners" were the largest and dominated groups who were primarily well-educated viewers and knew a lot about foreign culture. The other four groups – "escapees", "adherents", "adjusters", and "neutralists" – were subordinated.

Among *genre-related dimension* "(extreme) learners" and "neutralists" were dominated by all genres. This meant most participants either kept an open mind regarding learning cultural difference, or maintained a neutral attitude.

Among *genre-unrelated dimensions* "extreme learners" and "neutralists" dominated in "language" and "values" difference. This meant that in confronting both differences, more than half of the participants kept either a very open mind or a neutral attitude. "Extreme learners" and "adjusters" would like to learn or adjust their "way of life" and "education method" the most.

*(3) Impact of influential factors*

*All influential factors*:

- P*erception*: "American", "Asia", and "education" were the strongest predictors for the perception index. This indicated that if participants did not like American programs or liked Asian programs or had a higher education level, then they perceived a larger cultural difference than other viewers.

- *Acceptance*: "America", "Europe", and "knowledge" were the three strongest predictors for the acceptance index. This indicated that if participants liked American programs, or European programs, or knew a lot about foreign culture, then they accepted more cultural difference than other viewers.

- Generally, cultural proximity variables were the most influential predictors.

*Particular influential factors*:

- *Demographics*: Among demographic variables "education" remained the most important factor. "Gender", "age", and "income" influenced only a few dimensions.

- *Cultural exposure*: "Knowledge" and "viewing" were two significant predictors. A lot of knowledge was combined with a high perception degree and a high acceptance degree. A lot of viewing was not combined with a high perception degree, but with a high acceptance degree.

- *Cultural proximity*: Viewers liked American programs more than programs from other countries. A high preference for programs from culturally proximate countries (Asian) was combined with a low perception degree, but not with a low acceptance degree. A high preference for programs from culturally distant countries (America, Europe, and other non-Asian countries) was combined with high perception and high acceptance.

- *Regional difference*: Firstly, perception and acceptance indices of the five cities were significantly different to each other. Secondly, perception index was not significantly correlated to city ranking, whereas acceptance index was. Thus, viewers from more-open cities did not perceive more cultural difference than those from more-conservative cities, whereas they accepted greater difference than those from more-conservative cities.

# Part E: Conclusions

# 18 Conclusions

## 18.1 Summary

- **Theoretical framework**

*Culture* is a complex whole which includes values, beliefs, knowledge, art, morals, customs and habits, which are created over time and shared by members of a society. This framework of meanings allows the members of a society to understand each other and influence the ways in which people relate to one another.

*(1) Dimensions of cultural difference*

Cultural difference varies in quite different ways in people's lives:

Some scholars divided cultural differences into *horizontal dimensions*, e.g., Hofstede's five cultural-difference dimensions and Hall's concept of high- and low-context culture. In these concepts, each cultural-difference dimension has two poles to describe two opposite extreme situations. Each particular culture could be positioned somewhere within range between the two poles on a scale.

Some others divided cultural differences into *vertical levels*. For example, Hall's concept of cultural levels which considered hierarchical relationships between cultural dimensions.

Television programs also have cultural-difference dimensions if they are produced in one culture and broadcast in another. They are cultural goods as they say something, which must be understood and interpreted in a cultural context of television viewers.

*(2) Degree of cultural difference*

The starting point of this study concerns some controversial theoretical discussions about cultural difference in universal media products:

On the one hand, there is a belief in the world cultural convergence of universal media products. In this world, people genuinely understand each other. One result of cultural convergence is cultural imperialism, which has an overall effect of pushing cultural diversity more towards universal culture.

On the other hand, global flow of universal media products does by no means indicate universal reception and the end of difference. Despite surface appearances real differences remain and will probably stay that way for quite some time. Audiences could still feel cultural barriers, as they find it difficult to identify some unfamiliar or unexperienced culture in foreign media.

The essence of the discussions concerns the extent of cultural difference in universal media products. The term "intercultural adaptation" is used to describe an individual's cognitive developmental process and transformational learning to achieving an overall "fit" on cultural difference. The theoretical models about cultural adaptation are, for example, Lysgaard' U-Curve Model, Adler's Five Stages of Transitional Experience, and Bennett's Developmental Model of Intercultural Sensitivity.

The author developed a perception/acceptance matrix to describe the intercultural adaptation process of television viewers:

*a. Perception* indicates the degree to which viewers' own cultural frame is consistent with other cultural frames in foreign television programs.

*High vs. low perception*: The heterogeneity between cultures may look small or large in comparison. Cultural differences are embedded within similarities, and cultural similarities are embedded within differences.

*b. Acceptance* indicates the extent to which viewers accept and see the perceptual realities from other cultures as valid.

*High vs. low acceptance*: Some viewers react to cultural difference only with low acceptance and deep resistance. Some others are able not only to see cultural difference as rational, but also accept the fundamental premise that other cultures have.

c. *Perception/Acceptance Matrix*: Combined both (high/low) perception and (high/low) acceptance degree, the stages of cultural adaptation may vary across five areas. To shift from escapee to adjuster, viewers need an intercultural mindset to accumulate knowledge about foreign cultures.

- Escapees (high perception and low acceptance);
- Learners (high perception and high acceptance);
- Adherents (low perception and low acceptance);
- Adjusters (low perception and high acceptance);
- Neutralists (middle perception and middle acceptance).

- **Research methods**

In this study, China is a sample country used to examine the cultural difference in universal television programs, because firstly Chinese culture is a typical high-context culture compared to the low-context cultures of Western developed countries. Secondly, since the *Reform and Opening-Up Policy* was launched in 1978 China's society has changed radically and become more open to the outside world. Chinese viewers now have more opportunities to watch foreign television programs and thus must confront cultural difference. In contrast to many other studies in which the United States was considered as the center, this study examines cultural difference from the perspective of Chinese viewers in a high-context culture when they watch programs from low-context cultures.

The research object is cultural difference in foreign television programs in China from the perspective of Chinese television viewers. Three research questions were answered:

*(1) What types of foreign television programs are broadcast in China?*

*(2) What types of cultural difference are perceived by Chinese viewers?*

*(3) To what extent could they perceive and accept cultural difference? How large is the impact of influential factors?*

According to research questions author developed a three-stage methodological approach to data collection:

*(1) TV schedule analysis as a pre-study to gather television genres*

This pre-study aimed at firstly constructing an overview of the types of foreign television programs; secondly, at describing and evaluating some forms of cultural modification in foreign programs, which could reduce cultural barriers for viewers; thirdly, examining cultural proximity between the country of origin and China. This would provide a framework of the research object in this study. This part analyzed TV schedules from 37 television channels of eight television broadcasters in five sample cities in China.

*(2) Personal interview to gather cultural-difference dimensions*

Personal interviews with Chinese viewers aimed to firstly identify various cultural-difference dimensions which were interpreted by Chinese viewers. All of the identified dimensions were summarized in a catalogue in order to build an overview of the various cultural differences. Secondly, based on viewers' statements the cultural background of each dimension was explained.

In this part, a total of 36 interviews with Chinese television viewers were conducted in five sample cities. All interviewees were selected according to different demographics. However, due to the small number of participants, the representatives of the sample as well as comprehensiveness of cultural-difference dimensions could not be guaranteed.

*(3) Survey to evaluate degree of perception and acceptance*

Surveys with Chinese viewers aimed to firstly evaluate the perception and acceptance degree of each cultural-difference dimension quantitatively (including 10 dimension sets/42 cultural dimensions). All dimensions in the survey derived from the results of the television schedule analysis and the personal interviews. Each cultural dimension had both the perception index and the acceptance index on a five-point scale. The perception index indicated how large the cultural difference was that viewers perceived. The acceptance index indicated to what extent they accepted it. Secondly, all participants were divided into several viewer types according to both indices. Thirdly, the survey also examined the relationships between the perception/acceptance index and various influential factors. A total of 450 surveys were conducted in five sample cities.

- **Results 1: Genres of foreign television programs (TV schedule analysis)**

The results of the television schedule analysis indicated that generally foreign television programs accounted for a significant proportion of the total broadcasting time in China. Every sample channel had a certain time slot or slots for broadcasting foreign programs every week. The results consist of three parts:

*(1) Genres of foreign television programs*

Sports programs, television series, and documentaries were broadcast the most. Movies, music programs, and entertainment programs were broadcast less often. The other genres, such as news programs and talk shows, were not broadcast.

Themes of each genre were not broadcast at an equal level. The themes with a lower cultural content, such as action and science-fiction movies, sports programs about basketball and football, documentaries about nature and geography, as well as music programs about classical music were obviously broadcast more than other themes with greater cultural content.

*(2) Cultural modification*

For a variety of cultural, economic, or political reasons, the imported television programs must be modified in ways that make them more relevant and meaningful to Chinese viewers. Cultural modification included three aspects: content, language, and person modification. Each modification had three levels: minimal, medium, and maximal. The results demonstrated that the "content" of and "persons" in most foreign television program did not need to be modified, whereas the "language" of most programs had to be modified to a large degree.

*(3) Cultural proximity*

Countries of origin ranged from small (Asian countries) to large (U.S. and European countries) cultural proximity compared with China. Programs from the United States dominated in all genres, whereas the programs from culturally proximate Asian countries (e.g., Japan and Korea) accounted for only a small proportion of the total offering. Asian countries exported primarily television series to China. European countries concentrated their exporting on television series and sports programs.

In sum, the results of the television schedule analysis created an overview of universal television programs in China. In previous studies, these programs which were offered for the global market were often discussed as a whole and not distinguished with each other clearly and systematically. The catalogue of foreign universal programs in this study enables them to be investigated more structurally and makes a comparison between foreign universal programs much easier.

- **Results 2: Dimensions of cultural difference (personal interview)**

*(1) Types of cultural difference*

Based on numerous participants' statements about cultural difference in foreign television programs, a total of 10 dimension sets/42 dimensions were found and summarized in a catalogue.

- *Genre-related* cultural differences related to certain contents and formats of television programs which included six dimension sets – movies & television series, sports programs, documentaries, entertainment shows, news programs, and music programs.

- *Genre-unrelated* cultural differences did not relate to certain genres. They included four dimension sets: language, values, ways of life, and education methods.

The statements of Chinese participants indicated that to their mind there was a clear boundary between "our" and "their" cultures. They had two frameworks for interpreting own Chinese culture ("conservative") and cultures in foreign television programs ("open" and "liberal").

In sum, this catalogue of numerous types of cultural difference as well as the detailed interpretation of cultural background for each cultural dimension, builds an overview of differentiated types of cultural difference for different television genres. This offers a supplement to theories about cultural-difference dimensions in daily life which are not directly related to media products. This systematical catalogue made possible the quantitative evaluation and comparison across the whole range of possible dimensions in the survey. Without this concrete interpretation of particular dimensions, the discussions about cultural difference would have seemed over-generalized.

In contrast to numerous studies, which examined cultural difference in media products based on country level, this study addressed cultural difference from the perspective of audiences, because watching television occurs at an individual level. This audience perspective was also a complement to the cultural differences which were interpreted by media companies and the department of media censorship.

*(2) Types of Chinese television viewers*

All of the perceived cultural differences varied between large and small, while the degree of acceptance varied from not acceptable to totally acceptable. All participants were divided into five groups according to their degree of perception and acceptance:

- *Escapees* avoided confronting foreign culture and remained outsiders in regard to foreign programs. Cultural difference was high for them and they did not accept it.

- *Adherents* clearly did not perceive cultural difference, but they did not want to accept it and change their own culture.

- *Learners* considered cultural difference to be high, but they kept an open mind to learning it.

- *Adjusters* combined their own culture and foreign culture in mind. They adopted a new more integrative cultural frame which embraced some aspects of new cultures. They could not clearly distinguish cultural difference between their own and foreign cultures and accepted it totally.

- *Neutralists* neither cared about cultural difference, nor wanted to fit in with it.

The transition from escapees to adjusters is a long process based on the continual accumulation of intercultural experience. The starting point of transition was escapees. Adherents, learners, and neutralists were the middle stages. The end stage was adjusters.

In sum, these television viewer types according to their attitudes towards cultural difference made theoretical discussions about stages of intercultural adaptation more concrete and operational.

- **Results 3: Perception and acceptance of cultural difference (survey)**

*(1) Perception and acceptance of cultural difference*

*a. Cultural differences were large, but acceptable mostly.*

In the survey both the perception and the acceptance index for each cultural dimension made the quantitative examination of a whole range of various dimensions possible and reliable. The cut-off points of large/small as well as acceptable/unacceptable cultural difference were defined clearly on a 5-point scale.

The results of the survey demonstrated that the perception indices of all dimensions lay between "normal" and "large". The acceptance indices of all dimensions lay between "acceptable" and "normal". This finding indicated that although cultural differences were perceived to be large, viewers could mostly accept them.

This result answered the question of whether cultural convergence or cultural difference dominated in the minds of Chinese viewers. The answer was that *cultural difference* was still dominated by *perception* of cultural difference, whereas *cultural convergence* was dominated by *acceptance* of cultural difference:

- *Cultural difference dominated by perception*: The global flow of universal television programs did not indicate a universal perception of cultures. Chinese viewers perceived large cultural differences not only at surface level, but also at a deep level. A very strong Chinese identity still remained in the minds of almost all viewers, which clearly had not changed and could not be changed easily.

- *Cultural convergence dominated by acceptance*: Besides a high perception degree of cultural difference most viewers had begun to learn about cultural difference and had tried to accept it. Actually, these universal programs had attracted a large number of viewers in China, who had shared habits and tastes with similar market segments in other countries. In this borderless world viewers transcend different cultures and understand each other. Thus, the worries about cultural imperialism were not without foundation. It seemed clear that the values and ideologies in universal television programs from Western countries had significantly influenced the attitudes of Chinese viewers.

*b. Genres with little linguistic content were more acceptable.*

Music programs which contain little linguistic content were the most acceptable. In contrast, news programs in which speech was an important element had a lower acceptance degree.

This result partly supported the opinions of some scholars that programs which were not dependent on understanding the language sometimes attract potentially large audiences. Programs with high speech content, which caused a great deal of misunderstanding, were unpopular with foreign viewers.

*c. Language difference was the most unacceptable.*

Language difference was the most difficult to accept, especially because of inaccuracy of translation. Thus, most viewers preferred to watch foreign programs in the original language than dubbed.

This finding strongly supported that language difference was the most important factor in determining whether a cultural barrier existed when watching foreign programs. Language as a "natural" barrier had to be negotiated through dubbing or subtitling, which disrupted viewing pleasure.

*d. Education methods and way of life was the largest, but the most acceptable.*

Both dimensions "education methods" and "way of life" were perceived as the largest cultural difference; however, they were the most acceptable one. This finding indicated that Chinese viewers preferred the education methods and way of life they had seen depicted in foreign programs more than Chinese ones. Among the four aspects of education methods "imagination/creativity" was the largest, but also the most acceptable.

*e. Values was a large difference and unacceptable, especially religion.*

The cultural difference of "values" was the second most unacceptable dimension. Among the seven value dimensions "religion" difference was the largest and the most unacceptable.

The conclusions of *b, c, d,* and *e* indicated that the perception and acceptance degree of various cultural-difference types varied very strongly. The case for one dimension was not the case for another. Thus, an examination of cultural differences should always be differentiated instead of an over-generalization.

*(2) Stages of intercultural adaptation (viewer types)*

*a. Learners were the dominating viewer group.*

"(Extreme) learners" were the largest viewer group. Nearly half of the participants were "learners" or "extreme learners" who considered cultural difference to be large, but kept an open mind to learning about it. The other four viewer groups "escapees", "adherents", "adjusters", and "neutralist" were subordinated.

*b. Learners were well-educated and had a lot of intercultural knowledge.*

"(Extreme) learners" were primarily well-educated participants who possessed a great deal of knowledge about foreign cultures, whereas the participants with a low education level and little intercultural knowledge were primarily "escapees".

*c. Genre-related dimensions: domination of (extreme) learners and neutralists.*

Among the six genre-related dimensions both viewer groups "(extreme) learners" and "neutralists" dominated. This meant that most participants either kept an open mind towards learning about cultural difference, or kept a neutral attitude in order to avoid it.

*d. Genre-unrelated dimensions: domination of extreme learners and neutralists.*

Among the four genre-unrelated dimensions "extreme learners" and "neutralists" were dominated by "language" and "values" difference. This meant that in confronting both differences more than half of the participants either maintained a very open mind or a neutral attitude. "Extreme learners" and "adjusters" would like to learn or adjust "way of life" and "education method" the most.

On the whole, the domination of learners supported that the young generations in China who were born after China's *Reform and Opening-Up Policy* had learned a lot about Western cultures in a more open living environment compared to older generations. Only a small proportion of viewers kept away from foreign culture and adhered to Chinese culture.

*(3) Influential factors*

*a. Education was the strongest predictor among the four demographics.*

Among demographic variables "education" remained the most important factor. "Gender", "age", and "income" influenced only a few particular dimensions. The finding indicates that extent to which participants could perceive and accept cultural difference in foreign television programs depends primarily on what education level they have attained.

Thus, it was supported that education exerted a positive influence on the acceptance of cultural difference, because viewers with a high education had more opportunity to be in touch with foreign cultures through the media and had better intercultural mindsets allowing them to be receptive to something new from other cultures.

*b. "Knowledge" and "viewing" were two significant predictors.*

A high amount of knowledge was combined with a high perception degree and a high acceptance degree. This finding supported the proposal in the theoretical discussion that if television producers and viewers had a lot of knowledge in common, cultural difference was more acceptable for viewers. Their intercultural knowledge teaches them to use alternative cultural interpretations in order to understand things across cultural boundaries.

A high amount of viewing was *not* combined with a high perception degree, but with a high acceptance degree. This proved the positive influence of viewing on acceptance degree. Through a long process of watching foreign programs, viewers gradually accumulated intercultural experiences and were used to having a new way of thinking.

*c. Cultural proximity variables were the most influential predictors.*

Firstly, viewers liked American programs more than those from other countries in almost all genres. Only in the case of entertainment shows did viewers prefer the Asian ones.

This result proved how much viewers like foreign programs, which was related primarily to how successful the programs were in the world market (e.g., America movies), and not directly related to cultural proximity between exporting countries and China. Thus, some theories which assumed that viewers had a stronger preference and desire for television content originating from similar cultures that reflect their own language, cultural background, history, and values compared to content from culturally distant cultures, were not proved in this study.

Secondly, the preference for television program from culturally proximate countries (e.g., Japan and Korea) was a significant predictor for perception, but not for acceptance. A high preference for programs from culturally proximate countries (Asian) was combined with a low perception degree, but not a low acceptance degree.

Thirdly, the preference for television programs from culturally distant countries (America, Europe, and other non-Asian countries) was not only able to predict perception of some cultural dimensions, but also acceptance of all cultural dimensions very significantly and strongly. This meant that a high preference for programs from culturally distant countries was combined with both high perception and high acceptance.

*d. Regional difference was obvious. Acceptance was related to city ranking.*

Firstly, the perception and acceptance indices of the five cities were significantly different. Secondly, perception index was not significantly correlated to city ranking, whereas acceptance index was. Thus, viewers from more-open cities did not perceive more cultural difference than those from more-conservative ones, whereas they accepted more difference than those from more-conservative cities.

The results indicated that, on the one hand, regional differences still remained in China due to different regional cultures, degrees of economic development, and degrees of openness. Thus, viewers in more-open cities had a more open mind towards cultural difference than those in more-conservative cities. On the other hand, due to telecommunication advances, regional cultural differences were gradually being diminished with the rise of information-based attitude groups that share similar media-usage patterns.

*e. Perception: American, Asia, and education were the strongest predictors.*

American, Asia, and education were the strongest predictors for the perception index. This indicated that if participants did not like American programs, or liked Asian programs, or had a higher education level, then they perceived greater cultural difference than other viewers.

*f. Acceptance: America, Europe, and knowledge were the strongest predictors.*

America, Europe, and knowledge were the strongest predictors for the acceptance index. This indicated that if participants liked American programs, or European programs, or knew a lot about foreign culture, then they accepted more cultural difference than other viewers.

In Table 40 all results of hypotheses testing are summarized.

*Table 40: Summary of the Results of Hypotheses Testing*

| Hypothesis | | Results |
|---|---|---|
| Part I: Perception and acceptance of cultural difference | | |
| H1a | TV viewers perceive the cultural differences to be large. | Accepted |
| H1b | TV viewers accept the cultural differences mostly. | Accepted |

| | | |
|---|---|---|
| H2a | The genres with little linguistic content (music programs) have smaller cultural difference than those in which speech is an important element (news programs). | Rejected |
| H2b | The genres with little linguistic content (music programs) are more acceptable than those in which speech is an important element (news programs). | Accepted |
| H3a | Language difference is the difference that is the most difficult to accept. | Accepted |
| H3b | Viewers prefer watching foreign TV programs with dubbing rather than those in the original language with subtitles. | Rejected |
| H4a | The distance between the perception and acceptance index of education method is larger than other cultural dimensions. | Accepted |
| H4b | The distance between the perception and acceptance index of way of life is larger than other cultural dimensions. | Accepted |
| H5a | The cultural difference at a deep level is perceived to be larger than the cultural difference at superficial level. | Rejected |
| H5b | The cultural difference at a deep level is less acceptable than the cultural difference at a superficial level. | Rejected |
| **Part II: Viewer types** | | |
| H6a | "Learners" are the largest group of viewer types who perceive the cultural difference to be large, but they can accept it to a high degree. | Accepted |
| H6b | "Learners" are primarily well-educated viewers who know a great deal about foreign cultures. | Accepted |
| H7 | The viewer groups of "escapees", "adherents", and "adjusters" are smaller than "learners". | Accepted |
| H8 | For all viewer types the cultural difference in music programs is the smallest and the most acceptable. | Rejected |
| H9 | With language difference viewers are primarily "escapees" and "adherents". | Rejected |
| H10a | With cultural difference of way of life and education method, viewers are primarily "learners". | Accepted |
| H10b | "Learners" and "adjusters" prefer to learn and adjust their way of life and education method more than other dimensions. | (Partly) Accepted |
| **Part III: Influential factor** | | |
| H11a | Men perceive more cultural differences than women. | Rejected |
| H11b | Men accept more cultural differences than women. | Rejected |

| | | |
|---|---|---|
| H12a | Old viewers perceive more cultural difference than young viewers. | Rejected |
| H12b | Young viewers accept more cultural difference than old viewers. | Rejected |
| H13a | Viewers with a high level of education perceive more cultural difference than those with a low level of education. | Accepted |
| H13b | Viewers with a high level of education accept more cultural difference than those with a low level of education. | Accepted |
| H14a | Viewers with high incomes perceive more cultural difference than those with low income. | Rejected |
| H14b | Viewers with high incomes accept more cultural difference than those with low income. | Rejected |
| H15a | Viewers who know a lot about foreign culture perceive more cultural difference than others who know little. | Accepted |
| H15b | Viewers who know a lot about foreign culture accept more cultural difference than others who know little. | Accepted |
| H16a | Viewers who watch a lot of foreign TV programs perceive more cultural difference than those who watch little. | Rejected |
| H16b | Viewers who watch a lot of foreign TV programs accept more cultural difference than those who watch little. | Accepted |
| H17 | Viewers like TV programs from culturally proximate countries more than culturally distant countries. | Rejected |
| H18a | Viewers who like TV programs from culturally proximate countries perceive less cultural difference than those who do not. | Accepted |
| H18b | Viewers who like TV programs from culturally proximate countries accept less cultural difference than those who do not. | Rejected |
| H19a | Viewers who like TV programs from culturally distant countries perceive more cultural difference than those who do not. | (Partly) Accepted |
| H19b | Viewers who like TV programs from culturally distant countries accept more cultural difference than those who do not. | Accepted |
| H20a | Viewers from more open cities perceive more cultural differences than those from more conservative cities. | Rejected |
| H20b | Viewers from more open cities accept more cultural differences than those from more conservative cities. | Accepted |

Source: Own illustration.

- **Conclusion**

In China today, the foreign television programs from the United States, and European and Asian countries are a significant part of program schedules. Although these programs target international viewers, most must be culturally modified for Chinese audiences, especially in regards to language modification.

The empirical results indicated that most Chinese viewers still perceived a lot of large cultural differences from the surface to deep levels. However, most had an open mind towards learning about cultural difference and accepted it. This finding indicated that although Chinese viewers still drew a clear boundary between "their" as well as "our" cultures and Chinese cultural identity dominated in their minds, they had begun to try to see other cultures as rational and to integrate them into own cultural framework. Thus, most of them were still in the learning phase. Only a small portion of them had successfully adjusted cultural differences.

To what extent they perceived and accepted cultural difference was related to certain influential factors. Generally, the well-educated viewers who had a high level of intercultural knowledge or watched a lot of foreign programs were likely to accept greater cultural difference. The more they liked programs from culturally distant countries, the greater cultural difference they perceived and accepted. In contrast, the more they liked programs from culturally proximate countries, the less cultural difference they perceived. Viewers who lived in open cities had a more acceptable attitude towards foreign programs than those in conservative cities.

## 18.2 Discussion

This study has a number of limitations, some of which imply directions for needed research in the future:

*(1) Theoretical limitations*

*a. Enrich research perspectives*

Cultural difference in this study focused only on television viewers. It did not examine media corporations' attitudes towards cultural difference and their measures to reduce cultural difference in their television programs. Future research might request both television suppliers and receivers an assessment of cultural difference to examine whether they can really understand each other. Such studies might help to investigate whether cultural modification from the supplier side could actually help to reduce cultural difference as barriers from the receiver side. Viewers' attitudes towards foreign programming with and without cultural modification could be compared.

*b. Enrich media types*

Television was the only medium used to examine cultural-difference dimensions in this study, because television programs are the most popular and widespread foreign media content in China and it was the easiest accessible medium for viewers of different ages and in different regions compared to the internet, newspapers, and magazines. However, other foreign media in China would also be able to influence audiences' attitudes towards cultural difference. Additional studies could discuss cultural difference in other media areas and compare them in a theoretical framework.

*c. Discuss the translation of attitude into behavior*

The approach of cultural adaptation was unable to adequately explain the translation of attitude into behavior. Although these theoretical models interpret the sequence and stands of intercultural adaption, they did not seek to explain the mechanism of adaptation itself; in other words, how the attitudes towards intercultural adaptation translated into actual adaptive behavior. Further research could examine the relationship between attitudes and actions of adaptation. Discussion questions could be: How do viewers' "learn/adjust" or "escape/adherent" attitudes influence their viewing habits of foreign programs, or preferences for certain countries of origin? How do their attitudes influence their intercultural behavior in daily life? Future research should address these theoretical discussions.

*(2) Methodological implications*

*a. Concretize the questions in the survey*

The questions in the survey about perception/acceptance, knowledge, viewing, and cultural preference were not without limitations because they did not seem concrete enough. The answers from participants were based on their intuitive feelings. For example, they were not asked why they watched foreign programs, how they acquired intercultural knowledge (e.g., through media), of why they preferred certain programs from certain countries. These questions would help to delve deeper into the reasons for perception and acceptance. Furthermore, various European countries (e.g., Britten, Germany, and France) were considered as a whole "Europe" to evaluate cultural proximity. Actually, cultural difference between programming from these particular European countries might be large. In future studies, the survey questions should be asked in more detail.

*b. Regard a representative sample*

Due to a diverse and huge population in China, as well as limited financial resources, the representativeness of the samples in this study was generally not possible. The sample size was not large enough to present a representative result. The well-educated viewers and old viewers were under-represented. The difference between viewers in urban and rural areas was not considered, because the samples generated for this study were limited to five single metropolitan areas. Generalizing inferences from 450 samples in five cities to general populations in China was rather limited. Future research should engage more representative populations, in which different demographic groups are better distributed, in order to achieve more plausible results. Particularly, viewers from rural areas should have equal opportunity to viewers from urban areas. The difference between them might be very interesting.

*c. Enlarge the variance*

Several cautionary remarks need to be made here with regard to the interpretation and generalization of the findings. Although multiple control factors were used in the survey, the size of the variance of the sample, which was able to be explained, was relatively small. Eleven influential factors accounted for only 11.7% of variance in perception and 24.3% in acceptance. This indicated that there might be some other influential factors outside this work which significantly contributed to the perception and acceptance indices. Future research could add some new control factors to the analysis model which might be potentially important determinants.

*d. Enlarge the research period*

In this study personal interviews and surveys were carried out in 2009 and 2010. Thus, the results only explained viewers' attitudes towards cultural difference in this limited time period. Due to the quickly changing living environment and the development of foreign media in China, viewers' opinions might change very quickly. As the new post-90 generation grows older, the viewer groups of "learners" and "adjusters" might rise rapidly. The future study should observe the changes of viewers' opinions over a longer period.

*e. Increase participants' autonomy in the research*

Another limitation should be addressed is that the face-to-face interview or survey might have placed artificial restrictions on participants, which prevented them from being autonomous. Future studies could use experimental settings in which participants may have more autonomy to interpret their "original" attitudes towards cultural difference through their behavior and dialogue by watching certain foreign programs in the absence of unfamiliar people.

*(3) Empirical implications*

*a. Examine particular cultural-difference dimension in more detail*

This study is the first to find Chinese television viewers' overall attitudes towards cultural difference in foreign programs; thus, the investigation of particular cultural-difference dimensions is not detailed enough. Further research should examine certain dimensions in more detail and explain their reasons and performance at a deeper level.

*b. Enlarge the range of cultural-difference dimensions*

Due to the small sample size of participants for the personal interviews, the completeness of all cultural dimensions in the analysis model cannot be guaranteed. In this study, all 42 cultural dimensions were derived from personal interviews which were considered to be the most important in the opinion of Chinese viewers. No external cultural dimensions from other empirical studies or theories were used. Future studies could develop a more comprehensive catalogue of cultural-difference dimensions. If some new cultural dimensions are incorporated into the proposed model as variables, the reliably of the results will be increased substantially.

*c. Distinguish between cultural difference in television and daily life more clearly*

Notably, some cultural-difference dimensions in this study were an individual's perception made in daily life rather than features of television programs, especially several genre-unrelated cultural dimensions (e.g., values, way of life, and education method). The difficulty the author had in explaining cultural difference in television programs was most likely rooted in a general inability to distinguish whether cultural difference was derived from television programs or other places in daily life.

# Bibliography


Acheson, Keith; Maule, Christopher J. (2005): Understanding Hollywood's organisation and continuing success. In: John Sedgwick and Michael J. Pokorny (Eds.): An economic history of film. London: Routledge.

Acland, Charles R. (2003): Screen traffic: movies, multiplexes, and global culture. Durham: Duke University Press.

Adler, Peter S. (1975): The transitional experience. An alternative view of culture shock. In: *Journal of humanistic psychology* 15 (4), pp. 13–23.

Ajami, Riad A.; Goddard, G. Jason (2006): International business: theory and practice. New York: M.E. Sharpe.

Albarran, Alan B. (2006): Histrorical Trends and Patterns in Media Management Research. In: Alan B. Albarran, Sylvia M. Chan-Olmsted and Michael O. Wirth (Eds.): Handbook of media management and economics. New York: Erlbaum.

Albarran, Alan B.; Chan-Olmsted, Sylvia M. (Eds.) (1998): Global media economics. Commercialization, concentration and integration of world media markets. Ames: Iowa State University Press.

Albarran, Alan B.; Chan-Olmsted, Sylvia M.; Wirth, Michael O. (Eds.) (2006): Handbook of media management and economics. New York: Erlbaum.

Al-Shammari, Minwir (2010): Knowledge Management in Emerging Economies: Social, Organizational and Cultural Implementation. Hershey: Igi Global.

Altmeppen, Klaus-Dieter; Karmasin, Matthias (Eds.) (2003): Medien und Ökonomie. Band 1: Grundlagen der Medienökonomie: Kommunikations- und Medien-wissenschaft, Wirtschaftwissenschaft. Wiesbaden: Westdt. Verl.

Alvarado, Manuel (1996): Selling Television. In: Albert Moran (Eds.): Film policy. International, national and regional perspectives. London: Routledge, pp. 62–71.

Amant, Kirt St. (2007): Linguistic and cultural online communication issues in the global age. London: Information Science Reference.

Ambler, Tim; Xi, Chao; Witzel, Morgen (2008): Doing business in China. Oxon: Routledge.

Andersen, Peter A.; Hecht, Michael L. (2003): Nonverbal communication cross cultures. In: William B. Gudykunst (Eds.): Cross-cultural and intercultural communication. London: Sage Publications, pp. 73–90.

Arabski, Janusz; Wojtaszek, Adam (Eds.) (2011): Aspects of Culture in Second Language Acquisition and Foreign Language Learning. Berlin: Springer.

Asante, Molefi K.; Gudykunst, William B. (Eds.) (1989): Handbook of international and intercultural communication. Newbury Park: SAGE.

Bai, Ruoyun (2007): Anticorruption television dramas: between propaganda and popular culture in globalizing China. Illinois: University of Illinois at Urbana-Champaign.

Bao, Leiping (2004): China. In: Judith J. Slater (Eds.): Teen life in Asia. Westport: Greenwood Press, pp. 19–34.

Barker, Kathy (2002): At the helm: a laboratory navigator. New York: Cold Spring Harbor Laboratory Press.

Bednarek, Monika (2010): The language of fictional television: drama and identity. London: Continuum International Pub. Group.

Bell, Daniel A. (2008): Confucian political ethics. Princeton: Princeton University Press.

Bell, Daniel A. (2010): China's New Confucianism: Politics and Everyday Life in a Changing Society. Princeton: Princeton University Press.

Bennett, James; Strange, Niki (2011): Television as Digital Media. Durham: Duke University Press.

Bennett, Milton J. (2007): Basic concepts of intercultural communication: selected readings. Yarmouth: Intercultural Press.

Bennett, Milton J.; Castiglioni, Ida (2004): Embodied Ethnocentrism and the Feeling of Culture: A Key to Traning for Intercultural Competence. In: Dan Landies, Janet M. Bennett and Milton J. Bennett (Eds.): Handbook of intercultural training. Thousand Oaks: Sage Publications, pp. 249–265.

Bensinger, Viola (2005): Co-operation with China in the TV sector. Köln: Inst. für Rundfunkökonomie Univ. Köln (Arbeitspapiere des Instituts für Rundfunk-ökonomie an der Universität zu Köln, 207).

Berger, Arthur A. (Eds.) (1988): Media USA: process and effect. Munich: Longman.

Berger, Arthur A. (1992): Texts in Contexts: Analyzing Media and Popular Culture From a Cross-Cultural Perspective. In: Felipe Korzenny (Eds.): Mass media effects across cultures. Newbury Park: SAGE, pp. 11–22.

Berger, Arthur A. (2011): Media Analysis Techniques. Thousand Oaks: Sage Publications.

Berker, Thomas; Hartmann, Maren; Punie, Yves (Eds.) (2006): Domestication of media and technology. New York: Open University Press.

Björkstén, Johan; Hägglund, Anders (2010): How to Manage a Successful Business in China. Singapore: World Scientific.

Black, J. Stemart (1988): Work role transitions: A study of American expatriate managers in Japan. In: *Journal of International Business Studies* 1988 (19), pp. 277–294.

Bond, Michael H. (2010): The Oxford handbook of Chinese psychology. New York: Oxford University Press.

Bondebjerg, IB; Golding, Peter (2004): European culture and the media. Bristol: Intellect.

Brewer, Marilymn B.; Yuki, Masaki (2010): Culture and social identity. In: S. Kitayama and D. Cohen (Eds.): Handbook of Cultural Psychology. New York: Guilford Publications, pp. 307–322.

Bringle, Robert G.; Hatcher, Julie A.; Jones, Steven G. (2010): International service learning: conceptual frameworks and research. Sterling: Stylus.

Brodbeck, Felix C.; Chhokar, Jagdeep S.; House, Robert J. (2007): Culture and leadership in 25 societies: integration, conclusions, and future directions. In: Jagdeep S. Chhokar, Felix C. Brodbeck and Robert J. House (Eds.): Culture and leadership, across the world: The GLOBE Book of In-Depth Studies of 25 Societies. New York: Lawrence Erlbaum Associates, pp. 1023–1084.

Brownell, Susan; Wasserstrom, Jeffrey N. (2002): Chinese femininities, Chinese masculinities: a reader. London: University of California Press.

Bryant, Jennings; Vorderer, Peter (Eds.) (2006): Psychology of entertainment. Mahwah: Lawrence Erlbaum Associates.

Burnett, Robert (1996): The global jukebox. The international music industry. London: Routledge.

Butler, Jeremy G. (2007): Television: critical methods and applications. Mahwah: Lawrence Erlbaum Associates.

CCTV (2002): Year book of China Central Television 2002 (中国中央电视台年鉴 2002). Beijing: CCTV.

Chaffee, Steven H. (1992): Search for Change. Survey Studies of International Media Effects. In: Felipe Korzenny (Eds.): Mass media effects across cultures. Newbury Park: SAGE, pp. 35–54.

Chan-Olmsted, Sylvia M. (2006): Competitive strategy for media firms. Strategic and brand management in changing media markets. Mahwah: Lawrence Erlbaum Associates.

Chan-Olmsted, Sylvia M. (2006): Issues in Media Management and Technology. In: Alan B. Albarran, Sylvia M. Chan-Olmsted and Michael O. Wirth (Eds.): Handbook of media management and economics. New York: Erlbaum.

Chan-Olmsted, Sylvia M. (2006): Issues in Strategic Management. In: Alan B. Albarran, Sylvia M. Chan-Olmsted and Michael O. Wirth (Eds.): Handbook of media management and economics. New York: Erlbaum.

Chapel, William B. (1998): Effective management communication for China. In: Jan Selmer (Eds.): International management in China: cross-cultural issues. New York: Routledge, pp. 169–182.
Chaume, Frederic (2008): Teaching sychronisation in a dubbling course: Some didactic proposals. In: Jorge Díaz-Cintas (Eds.): The didactics of audiovisual translation. Amsterdam: John Benjamins Pub. Co, pp. 129–140.
Cheng, Winnie (2003): Intercultural conversation. Amsterdam: John Benjamins Pub.
Cheong, Weng E. (1997): The Hong merchants of Canton: Chinese merchants in Sino-Western trade. Richmond: Curzon.
Chhokar, Jagdeep S.; Brodbeck, Felix C.; House, Robert J. (Eds.) (2007): Culture and leadership, across the world: The GLOBE Book of In-Depth Studies of 25 Societies. New York: Lawrence Erlbaum Associates.
Chiaro, Delia (Eds.) (2010): Translation, humor and the media. London: Continuum.
Chimombo, Moira; Roseberry, Robert L. (1998): The power of discourse: an introduction to discourse analysis. Mahwah: Lawrence Erlbaum Associates.
Chu, C.Y. Cyrus; Yu, Ruoh-Rong (2010): Understanding Chinese families: a comparative study of Taiwan and Southeast China. New York: Oxford University Press.
Cohen, Akiba A.; Roeh, Itzhak (1992): When Fiction and News Cross Over the Border. Notes on Differential Readings and Effects. In: Felipe Korzenny (Eds.): Mass media effects across cultures. Newbury Park: SAGE, pp. 23–34.
Cohen, Jacob (1988): Statistical power analysis for the behavioral sciences. Hillsdale: Erlbaum.
Collins, R. (1998): From satellite to single market: new communication technology and European Public Service television. New York: Routledge.
Collins, Richard (1990): Culture, communication, and national identity: the case of Canadian television. Toronto: University of Toronto Press.
Collins, Richard (1990): Television: policy and culture. London: Routledge.
Corner, John; Harvey, Sylvia (1991): Enterprise and heritage: crosscurrents of national culture. London: Routledge.
Creeber, Glen (Eds.) (2007): The television genre book. British Film Institute. London: British Film Institute.
Croteau, David; Hoynes, William (2001): The business of media. Corporate media and the public interest. Thousand Oaks: Pine Forge Press.
CVSC-Sofres Media (2005): China Television Annals. Beijing: CVSC-Sofres Media.

Damm, Jens (2001): WorldWideWeb in China and Taiwan. In: *Berliner China-Hefte* 2001 (20).

Deardorff, Darla K. (2009): The Sage handbook of intercultural competence. Thousand Oaks: Sage Publications.

Derr, Clyde B.; Roussillon, Sylvia; Bournois, Frank (Eds.) (2002): Cross-cultural approaches to leadership development. Westport: Quorum Books.

Devereux, Eoin (2007): Understanding the media. London: Sage Publications.

Díaz-Cintas, Jorge (Eds.) (2008): The didactics of audiovisual translation. Amsterdam: John Benjamins Pub. Co.

Donald, Stéphanie; Keane, Michael; Hong, Yin (2002): Media in China: consumption, content and crisis. London: Routledge Curzon.

Doyle, Gillian (2002): Understanding media economics. London: SAGE.

Dracklé, Dorle (2005): Vergleichende Medienethnografie. In: Andreas Hepp (Eds.): Globalisierung der Medienkommunikation. Eine Einführung. Wiesbaden: VS Verl. für Sozialwiss.

Drenth, Pieter J.D.; Thierry, Henk; Wolff, Charles J.de (2001): A Handbook of Work and Organizational Psychology. Hove: Psychology Press.

Earl, Peter E.; Kemp, Simon (1999): The Elgar companion to consumer research and economic psychology. Cheltenham: Edward Elgar.

Easterby-Smith, Mark; Lyles, Marjorie A. (2005): The Blackwell handbook of organizational learning and knowledge management. Malden: Blackwell.

Eickhoff, Melanie; Hutt, Bettina (2004): Gegenwärtige Ausgestaltung und wünschenswerte Fortentwicklung der Fernsehordnung der Volksrepublik China. Am Beispiel von China central television (CCTV). Köln: Inst. für Rundfunkökonomie (Arbeitspapiere des Instituts für Rundfunkökonomie an der Universität zu Köln, 188).

Feng, Zhuo (2011): Das zweigleisige Fernsehsystem in der Volksrepublik China. Propaganda und Kommerz. München: AVM.

Fischer, Doris (2001): Rückzug des Staates aus dem chinesischen Mediensektor? Neue institutionelle Arrangements am Beispiel des Zeitungsmarktes. Duisburg: Universität/Forschungsinstitut für wirtschaftliche Entwicklung (Duisburger Arbeitspapiere zur Ostasienwirtschaft, 56).

Fischer, Doris (2002): Stability and commercial information. An economic approach and a look at China. In: *Asien - Hamburg* (84), pp. 69–76.

Fischer, Doris (2003): Understanding marketisation within the Chinese information sector. Taking television as an example. Köln: Inst. für Rundfunkökonomie Univ. Köln (Arbeitspapiere des Instituts für Rundfunkökonomie an der Universität zu Köln, 175).

Fiske, John (2002): Augenblicke des Fernsehens. In: Claus Pias (Eds.): Kursbuch Medienkultur. Die maßgeblichen Theorien von Brecht bis Baudrillard. Stuttgart: DVA.

Fiske, John (2002): Introduction to Communication Studies. London: Taylor and Francis.

Fu, W. Wayne; Sim, Clarice (2010): Examining International Country-to-Country Flow of Theatrical Films. In: *Journal of Communication* 60 (1), pp. 120–143.

Fu, Wei-Jen W.; Lee, Tracy K. (2008): Economic and cultural influences on the theatrical consumption of foreign films in Singapore. In: *The journal of media economics* 21 (1), pp. 1–27.

Furnham, Adrian; Bochner, Stephen (1986): Culture shock: psychological reactions to unfamiliar environments. London: Methuen.

Gershon, Richard A. (2006): Issues in Transnational Media Management. In: Alan B. Albarran, Sylvia M. Chan-Olmsted and Michael O. Wirth (Eds.): Handbook of media management and economics. New York: Erlbaum.

Gifford, Rob (2008): China Road: One Man's Journey Into the Heart of Modern China. London: BLOOMSBURY PUBLISHING.

Gillespie, Kate; Jeannet, Jean-Pierre; Hennessey, H. David (2010): Global Marketing. Mason: Cengage Learning.

Ginsburgh, Victor A.; Throsby, David (2006): Handbook of the economics of art and culture. Amsterdam: Elsevier North-Holland.

Glaister, Keith W.; Husan, Rumy H.; Buckley, Peter J. (2004): Strategic business alliances: an examination of the core dimensions. Cheltenham: Edward Elgar.

Goff, Patricia M. (2007): Limits to liberalization: local culture in a global marketplace. Ithca: Cornell University Press.

Gonzalez, Christiane F. (1989): Translation. In: Molefi K. Asante and William B. Gudykunst (Eds.): Handbook of international and intercultural communication. 2. printing. Newbury Park: SAGE, pp. 484–501.

Gonzalez, Virginia (2004): Second language learning: cultural adaptation processes in international graduate students in U.S. universities. Lanham: University Press of America.

Gopal, Sarvepalli (2008): History of humanity: scientific and cultural development. The twentieth century. London: Routledge.

Gozzi, Raymond (1992): Mass Media Effects in High- and Low-Context Cultures. In: Felipe Korzenny (Eds.): Mass media effects across cultures. Newbury Park: SAGE, pp. 55–66.

Grisold, Andrea (2004): Kulturindustrie Fernsehen. Zum Wechselverhältnis von Ökonomie und Massenmedien. Wien: Löcker.

Gudykunst, W.B (2005): Theorizing about intercultural communication. Thousand Oaks: SAGE.

Gudykunst, W.B; Ting-Toomey, S.; Nishida, T. (Eds.) (1996): Communication in personal relationships across cultures. Thousand Oaks: Sage Publications.

Gudykunst, William B. (Eds.) (2003): Cross-cultural and intercultural communication. London: Sage Publications.

Gullahorn, Jeanne E. (1963): An extension of the U-curve hypothesis. In: *Journal of Social Issues* 19 (3), pp. 33–47.

Gundling, Ernest; Zanchettin, Anita; Aperian Global (2007): Global diversity: winning customers and engaging employees within world markets. Boston: Nicholas Brealey International.

Guo, Zhenzhi (2004): WTO, "Chanye Hua" of the media and chinese television. Köln: Verein zur Förderung der Forschung auf dem Gebiet der Rundfunkökonomie.

Guthrie, Doug (2009): China and globalization: the social, economic and political transformation of Chinese society. New York: Routledge.

Hall, Edward T. (1977): Beyond culture. Garden City: Anchor Pr./Doubleday.

Hall, Edward T.; Hall, Mildred R. (1990): Understanding cultural differences. Germans, French and Americans. Yarmouth: Intercultural Press.

Hallin, Daniel C. (1994): We keep America on top of the world: television journalism and the public sphere. London: Routledge.

Hamilton, Cheryl (2010): Communicating for Results: A Guide for Business and the Professions. Wadsworth: Cengage Learning.

Hatim, Basil; Munday, Jeremy (2004): Translation: an advanced resource book. London: Routledge.

Heinrich, Jürgen (1994): Medienökonomie. Band 1: Mediensystem, Zeitung, Zeitschrift, Anzeigenblatt. Opladen: Westdt. Verl.

Hepp, Andreas (2004): Netzwerke der Medien. Medienkulturen und Globalisierung. Wiesbaden: VS Verlag für Sozialwissenschaften.

Hepp, Andreas (Eds.) (2005): Globalisierung der Medienkommunikation. Eine Einführung. Wiesbaden: VS Verl. für Sozialwiss.

Hillebrand, Jens (2008): Cross-Cultural Competence: Analysis of a Sino-Western Negotiation Setting. Munich: GRIN Verlag GmbH.

Hofstede, Geert (1991): Culture's consequences. International differences in work-related values. Newbury Park: SAGE.

Hofstede, Geert H. (2001): Culture's consequences: comparing values, behaviors, institutions, and organizations across nations. Thousand Oaks: SAGE.

Hong, Junhao (1998): The internationalization of television in China: the evolution of ideology, society, and media since the reform. Westport: Praeger.

Hoskins, Colin; McFadyen, Stuart; Finn, Adam (1997): Global television and film. An introduction to the economics of the business. Oxford: Clarendon Press.

Hoskins, Colin; McFadyen, Stuart; Finn, Adam (2004): Media economics. Applying economics to new and traditional media. Thousand Oaks: SAGE.

Hoskins, Colin; Mirus, Rolf (1988): Reasons for the US dominance of the international trade in television programmes. In: *Media, Culture and Society* 10 (4), pp. 499–515.

Howell, Jude (Eds.) (2004): Governance in China. Lanham: Rowman & Littlefield Publishers.

Hua, Gao (2011): Food augmentation methods and food substitutes during the great famine. In: Felix Wemheuer and Kimberley E. Manning (Eds.): Eating bitterness: new perspectives on China's Great Leap Forward and famine. Vancouver: UBC Press, pp. 171–196.

Huang, Quanyu; Andrulis, Richard S.; Chen, Tong (1994): A guide to successful business relations with the Chinese: opening the Great Wall's gate. Binghamton: International Business Press.

Hunter, Gordon; Tan, Felix (2005): Advanced Topics in Global Information Management. Hershey: Igi Global.

Iwabuchi, Kōichi (2002): Recentering globalization: popular culture and Japanese transnationalism. Durham: Duke University Press.

Jandt, Fred E. (1998): Intercultural communication. An introduction. Thousand Oaks: SAGE.

Jandt, Fred E. (2009): An Introduction to Intercultural Communication: Identities in a Global Community. Thousand Oaks: SAGE.

Jarren, Otfried; Donges, Patrick (2002): Politische Kommunikation in der Mediengesellschaft. Eine Einführung. Band 2: Akteure, Prozesse und Inhalte. Wiesbaden: Westdeutscher Verlag.

Jedynak, Malgorzata (2011): Attitudes of English teachers towards developing intercultural communicative competence. In: Janusz Arabski and Adam Wojtaszek (Eds.): Aspects of Culture in Second Language Acquisition and Foreign Language Learning. Berlin: Springer, pp. 63–76.

Jia, Wenshan; Lu, Xing; Heisey, D. Ray (2002): Chinese communication theory and research: reflections, new frontiers, and new directions. Westport: Ablex Pub.

Jing, Qicheng; Zuo, Qihua (2000): Mental health problems of Chinese school children. In: Nirbhay N. Singh (Eds.): International perspectives on child & adolescent mental health: selected proceedings of the First International

Conference on Child & Adolescent Mental Health, Hong Kong, June 1998. Kidlington: Elsevier, pp. 1–20.

Jolley, Richard P. (2010): Children and pictures: drawing and understanding. Malden: Wiley-Blackwell.

Keane, Michael (2004): A revolution in television and a great leap forward for innovation? China in the global television format business. In: Albert Moran and Michael Keane (Eds.): Television Across Asia: Television Industries, Programme Formats and Globalization. New York: Routledge, pp. 88–104.

Keane, Michael (2004): Bringing culture back in. In: Jude Howell (Eds.): Governance in China. Lanham: Rowman & Littlefield Publishers, pp. 77–96.

Keane, Michael; Fung, Anthony Y.H.; Moran, Albert (2007): New television, globalisation, and the East Asian cultural imagination. Hong Kong: Hong Kong University Press.

Keegan, Warren J.; Schlegelmilch, Bodo B. (2001): Global marketing management: a European perspective. Harlow: Financial Times Prentice Hall.

Keillor, Bruce D.; Kannan, Vijay R. (Eds.) (2011): International Business in the 21st Century. Santa Barbara: Greenwood Pub Group.

Kendall, Diana (2011): Sociology in Our Times: The Essentials. Belmont: Cengage Learning.

Kim, Young Y. (Eds.) (1988): Cross cultural adaptation. Current approaches. Newberry Park: SAGE.

Kitayama, S.; Cohen, D. (Eds.) (2010): Handbook of Cultural Psychology. New York: Guilford Publications.

Kitayama, Shinobu; Cohen, Dov (2010): Handbook of Cultural Psychology. New York: Guilford Publications.

Kleinman, Arthur; Yunxiang, Yan; Jing, Jun (2011): Deep China: The Moral Life of the Person, What Anthropology and Psychiatry Tell Us about China Today. Los Angeles: University of California Press.

Kloss, Ingomar; Abé, Makoto (Eds.) (2001): Advertising worldwide: advertising conditions in selected countries. Stralsund: Springer.

Kluckhohn, Florence R.; Strodtbeck, Fred L. (1973): Variations in value orientations. Westport, Conn: Greenwood Press.

Knight, Nick (2005): Marxist philosophy in China: from Qu Qiubai to Mao Zedong, 1923-1945. Dordrecht: Springer.

Korzenny, Felipe (Eds.) (1992): Mass media effects across cultures. Newbury Park: SAGE.

Krizan, A. C.; Patricia, Merrier; Logan, Joyce P. (2007): Business Communication. Mason: Cengage Learning.

Kuhn, Robert L. (2009): How China's leaders think: the inside story of China's reform and what this means for the future. Singapore: John Wiley & Sons.

Küng, Lucy (2009): Strategic management in the media. From theory to practice. Los Angeles: SAGE.

Landies, Dan; Bennett, Janet M.; Bennett, Milton J. (Eds.) (2004): Handbook of intercultural training. Thousand Oaks: Sage Publications.

Latham, Kevin (2001): Between markets and mandarins: journalists and the rhetorics of transition in southern China. In: Brian Moeran (Eds.): Asian media productions. Honolulu: University of Hawai'i Press, pp. 89–107.

Latzer, Michael (2002): Selbst- und Ko-Regulierung im Mediamatiksektor. Alternative Regulierungsformen zwischen Staat und Markt. Wiesbaden: Westdt. Verl.

Lau, Sing (1996): Growing up the Chinese way: Chinese child and adolescent development. Hongkong: Chinese University Press.

Lee, Francis L.F. (2006): Cultural discount and cross-culture predictability. Examing the box office performance of American movies in Hong Kong. In: *The journal of media economics* 19 (4), pp. 259–278.

Lee, Francis L. F. (2009): Cultural discount of cinematic achievement. The academy awards and US movies' East Asian box office. In: *Journal of cultural economics* 33 (4), pp. 239–263.

Lewin, Manuel (2004): Der Beitritt Chinas zur WTO und seine Auswirkungen (The influence of the China's WTO accession). Basel: WWZ of Basel University.

Li, Gucheng (1995): A glossary of political terms of the People's Republic of China. Hongkong: Chinese University Press.

Lim, Sunsun (2006): From culture to information revolution. ICT domestication by middle-class Chinese families. In: Thomas Berker, Maren Hartmann and Yves Punie (Eds.): Domestication of media and technology. New York: Open University Press, pp. 185–201.

Liu, Hong (2001): The market strategies of Chinese media (中国传媒的市场对策). Beijing: Beijing Broadcasting University.

Luhmann, Niklas (2002): Die Unwahrscheinlichkeit der Kommunikation. In: Claus Pias (Eds.): Kursbuch Medienkultur. Die maßgeblichen Theorien von Brecht bis Baudrillard. Stuttgart: DVA, pp. 55–66.

Lysgaard, Sverre (1955): Adjustment in a foreign society. Norwegian Fulbright grantees visiting the United States. In: *International Social Science Bulletin* 7, pp. 45–51.

Mao, Di (2008): A brief history of Chinese situation comedies. In: Ying Zhu, Michael Keane and Ruoyun Bai (Eds.): TV drama in China. Hong Kong: Hong Kong University Press, pp. 117–128.

Martin, Judith N.; Nakayama, Thomas K. (2009): Intercultural Communication in Contexts. New York: McGraw-Hill Higher Education.

Martinelli, Silvio; Gillert, Arne (2003): Intercultural learning T-kit. Strasbourg: Council of Europe.

Massing, Herta H. (1988): "Dallas" among German viewers. In: Arthur A. Berger (Eds.): Media USA: process and effect. Munich: Longman, pp. 194.

Mauk, David; Oakland, John (2005): American civilization: an introduction. Oxon: Routledge.

McAnany, Emile G.; Wilkinson, Kenton T. (1996): Mass media and free trade: NAFTA and the cultural industries. Austin: University of Texas Press.

McIntyre, Steve (1996): Art and industry. Regional film and video policy in the UK. In: Albert Moran (Eds.): Film policy. International, national and regional perspectives. London: Routledge.

McKenzie, Robert (2006): Comparing media from around the world. London: Pearson/Allyn and Bacon.

Miller, Katherine (2008): Organizational communication: approaches and processes. Wadsworth: Cengage Learning.

Mindess, Anna (2006): Reading between the signs: intercultural communication for sign language interpreters. Boston: Intercultural Press.

Mishan, Freda (2005): Designing authenticity into language learning materials. Bristol: Intellect.

Mitchell, Matthew C.; Alon, Ilan; Herbert, Theodor T. (2011): Assessing and managing political risk. In: Bruce D. Keillor and Vijay R. Kannan (Eds.): International Business in the 21st Century. Santa Barbara: Greenwood Pub Group, pp. 107–136.

Moeran, Brian (Eds.) (2001): Asian media productions. Honolulu: University of Hawai'i Press.

Moody, Peter R. (2007): Conservative thought in contemporary China. Plymouth: Rowman & Littlefield.

Mooij, Marieke de (2001): The impact of culture on advertising. In: Ingomar Kloss and Makoto Abé (Eds.): Advertising worldwide: advertising conditions in selected countries. Stralsund: Springer, pp. 1–22.

Mooij, Marieke K. de (2009): Global Marketing and Advertising: Understanding Cultural Paradoxes. Thousand Oaks: SAGE.

Moran, Albert (Eds.) (1996): Film policy. International, national and regional perspectives. London: Routledge.

Moran, Albert (1996): Terms for a Reader. Film, Hollywood, national cinema, cutlural identity and film policy. In: Albert Moran (Eds.): Film policy. International, national and regional perspectives. London: Routledge, pp. 1–22.

Moran, Albert; Keane, Michael (Eds.) (2004): Television Across Asia: Television Industries, Programme Formats and Globalization. New York: Routledge.

Moran, Robert T.; Harris, Philip R.; Moran, Sarah V. (2010): Managing Cultural Differences: Global Leadership Strategies for Cross-Cultural Business Success. Burlington: Butterworth Heinemann.

Morley, David (2002): Television, audiences and cultural studies. London: Routledge.

Morley, David; Robins, Kevin (1997): Spaces of identity. Global media, electronic landscapes and cultural boundaries. London: Routledge.

Mou, Sherry J. (1999): Presence and presentation: women in the Chinese literati tradition. New York: St. Martin's Press.

Napoli, Philip M. (2004): The Impact of Regulatory Change on Media Market Competition and Media Management. Mahwah: Lawrence Erlbaum Associates.

Narula, Uma (2006): Communication Models. New Delhi: Atlantic Publishers & Distributors.

Nauck, Bernhard; Steinbach, Anja (Eds.) (2005): Generatives Verhalten und Generationenbeziehungen: Festschrift für Bernhard Nauck zum 60. Geburtstag. Wiesbaden: VS Verlag für Sozialwissenschaften.

NBSC (National Bureau of Statistics of China) (2011): The main statistic report of Chinese population allocation in 2010 (2010年第六次全国人口普查主要数据公报). Available at: http://www.stats.gov.cn/tjgb/rkpcgb/qgrkpcgb/t20110428_ 402722232.htm, accessed November 10, 2011.

Neale, Steve (2007): Genre and television. In: Glen Creeber (Eds.): The television genre book. London: British Film Institute.

Neale, Steve (2007): Studying Genre. In: Glen Creeber (Eds.): The television genre book. London: British Film Institute.

Niemeier, Susanne; Dirven, René (1997): The language of emotions: conceptualization, expression, and theoretical foundation. Philadelphia: John Benjamins.

Noam, Eli M. (1993): Media Americanization, National Culture, and Forces of Integration. In: Eli M. Noam and Joel C. Millonzi (Eds.): The International market in film and television programs. Norwood: Ablex Publ. Corp, pp. 41–58.

Noam, Eli M.; Millonzi, Joel C. (Eds.) (1993): The International market in film and television programs. Norwood: Ablex Publ. Corp.

Nolan, Riall W. (1999): Communicating and adapting across cultures. Living and working in the global village. Westport: Bergin & Garvey.

Oberhofer, Tom (1989): The changing cultural discount rate. In: *Review of social economy* 47 (1), pp. 43–54.
OECD (2002): China in the Global Economy Foreign Direct Investment in China: Challenges and Prospects for Regional Development. Paris: Éditions OCDE.
OECD (2010): Tackling Inequalities in Brazil, China, India and South Africa: The Role of Labour Market and Social Policies. Paris: Organisation for Economic Co-operation and Development.
Oh, Jeongho (2001): International trade in film and the self-sufficiency ratio. In: *The journal of media economics* 14 (1), pp. 31–44.
Oliver, Marry B.; Kim, Jinkee; Sanders, Meghan S. (2006): Personality. In: Jennings Bryant and Peter Vorderer (Eds.): Psychology of entertainment. Mahwah: Lawrence Erlbaum Associates, pp. 329–337.
Ollig, Stefan (2002): Rahmenbedingungen und Entwicklungsmöglichkeiten ausländischer Fernsehprogrammanbieter in der VR China. Köln: Inst. für Rundfunkökonomie Univ. Köln (Arbeitspapiere des Instituts für Rundfunk-ökonomie an der Universität zu Köln, 156).
Parsons, Talcott; Shils, Edward; Smelser, Neil J. (2001): Toward a general theory of action: theoretical foundations for the social sciences. New Brunswick: Transaction Publishers.
Pedersen, Paul (1995): The five stages of culture shock: critical incidents around the world. Westport: Greenwood Press.
Pei, Minxin (1998): From Reform to Revolution: The Demise of Communism in China and the Soviet Union. Boston: Harvard University Press.
Petty, Sheila; Sherbert, Garry; Gérin, Annie (2005): Canadian cultural poesis: essays on Canadian culture. Waterloo: Wilfrid Laurier University Press.
Phalen, Patricia F. (2006): Audience Research and Analysis. In: Alan B. Albarran, Sylvia M. Chan-Olmsted and Michael O. Wirth (Eds.): Handbook of media management and economics. New York: Erlbaum.
Pias, Claus (Eds.) (2002): Kursbuch Medienkultur. Die maßgeblichen Theorien von Brecht bis Baudrillard. Stuttgart: DVA.
Pickton, David; Masterson, Rosalind (2010): Marketing: An Introduction. London: Sage Publications.
Poyatos, Fernando (Eds.) (1997): Nonverbal communication and translation: new perspectives and challenges in literature, interpretation and the media. Amsterdam: J. Benjamins.
Qian, Gong (2008): A trip down memory lane: remaking and rereading the red classics. In: Ying Zhu, Michael Keane and Ruoyun Bai (Eds.): TV drama in China. Hong Kong: Hong Kong University Press, pp. 157–172.

Reca, Ángel A. (2006): Issues in Media Product Management. In: Alan B. Albarran, Sylvia M. Chan-Olmsted and Michael O. Wirth (Eds.): Handbook of media management and economics. New York: Erlbaum.

Rohn, Ulrike (2010): Cultural barriers to the success of foreign media content. Western Media in China, India, and Japan. Frankfurt am Main: Peter Lang.

Rohn, Ulrike (2011): Lacuna or Universal? Introducing a new model for understanding crosscultural audience demand. In: *Media, Culture & Society* 33 (4), pp. 631–641.

Sadri, Houman A.; Flammia, Madelyn (2011): Intercultural Communication: A New Approach to International Relations and Global Challenges. London: Continuum.

Salmon, Dirk (2010): Expatriation in China. Munich: GRIN Verlag GmbH.

Samovar, Larry A.; Porter, Richard E.; McDaniel, Edwin R. (2009): Communication Between Cultures. Boston: Cengage Learning.

Samovar, Larry A.; Porter, Richard E.; McDaniel, Edwin R. (2011): Intercultural Communication: A Reader. Boston: Cengage Learning.

SARFT (2002): Year book of Chinese television 2002. Beijing: SARFT.

SARFT (2006): Guideline of SARFT. Beijing: SARFT. Available at: http://www. sarft.gov.cn/downstage/page_3.jsp, accessed September 03, 2009.

Sargent, Lyman T. (2008): Contemporary Political Ideologies: A Comparative Analysis. Belmont: Wadsworth.

Saxon, Mike (2006): An American's guide to doing business in China: negotiating contracts and agreements, understanding culture and customs, marketing products and services. Avon: Adams Media.

Schmidt, Wallace V.; Conaway, Roger N.; Easton, Susan S.; Wardrope, William J. (2007): Communicating globally. Intercultural communication and international business. Thousand Oaks: SAGE.

Schopenhauer, Arthur (1992): On language and words. In: Rainer Schulte and John Biguenet (Eds.): Theories of translation: an anthology of essays from Dryden to Derrida. London: University of Chicago Press, pp. 32–35.

Schulte, Rainer; Biguenet, John (Eds.) (1992): Theories of translation: an anthology of essays from Dryden to Derrida. London: University of Chicago Press.

Sedgwick, John; Pokorny, Michael J. (Eds.) (2005): An economic history of film. London: Routledge.

Selmer, Jan (Eds.) (1998): International management in China: cross-cultural issues. New York: Routledge.

Shaules, Joseph (2007): Deep culture: the hidden challenges of global living. Clevedon: Multilingual matters.
Shen, Qingsong; Shun, Kwong-loi (2007): Confucian ethics in retrospect and prospect. Washington DC: Council for Research in Values and Philosophy.
Sheng, Xuewen (2005): Families and intergenerational relationships in China. In: Bernhard Nauck and Anja Steinbach (Eds.): Generatives Verhalten und Generationenbeziehungen: Festschrift für Bernhard Nauck zum 60. Geburtstag. Wiesbaden: VS Verlag für Sozialwissenschaften, pp. 218–229.
Shirk, Susan L. (2011): Changing media, changing China. New York: Oxford University Press.
Singh, Nirbhay N. (Eds.) (2000): International perspectives on child & adolescent mental health: selected proceedings of the First International Conference on Child & Adolescent Mental Health, Hong Kong, June 1998. Kidlington: Elsevier.
Sklan, Carole (1996): Peripheral visions. Regionalism, Nationalism, Internationalism. In: Albert Moran (Eds.): Film policy. International, national and regional perspectives. London: Routledge.
Slater, Judith J. (Eds.) (2004): Teen life in Asia. Westport: Greenwood Press.
Son, Junghwa (2007): Indian consumer purchase behavior of foreign brand jeans. Ann Arbor: Oklahoma State University.
Sora, Park (2004): China's consumption of Korean television dramas. An empirical test of the "cultural discount" concept. In: *Korea Journal* 44 (4), pp. 265–290.
Staiger, Brunhild (2003): Das große China-Lexikon. Geschichte, Geographie, Gesellschaft, Politik, Wirtschaft, Bildung, Wissenschaft, Kultur: eine Veröffentlichung des Instituts für Asienkunde Hamburg. Darmstadt: Wiss. Buchges.
Steinberg, Sheila (2007): An Introduction to Communication Studies. Wetton: Juta.
Stewart, Edward C.; Bennett, Milton J. (1991): American cultural patterns: a cross-cultural perspective. Yarmouth: Intercultural Press.
Straubhaar, Joseph D.; Greenberg, Bradley S.; Ferreira, Leonardo (1992): What Makes News. Western, Socialist, and Third-World Television Newscasts Compared in Eight Countries. In: Felipe Korzenny (Eds.): Mass media effects across cultures. Newbury Park: SAGE, pp. 89–119.
Tian, Qing (2004): A transcultural study of ethical perceptions and judgements between Chinese and German businessmen. Munich: M-Press.

Ting-Toomey, Stella (1999): Communicating across cultures. New York: Guilford Press.
To, Yiu-Ming (1998): The Marketization of Media in China: Economic Freewheeling in a Political Straitjacket. In: Alan B. Albarran and Sylvia M. Chan-Olmsted (Eds.): Global media economics. Commercialization, concentration and integration of world media markets. Ames: Iowa State University Press, pp. 266–283.
Tonkin, Humphrey; Frank, Maria E. (2010): The Translator As Mediator of Cultures. Amsterdam: John Benjamins Pub.
Torbiörn, Ingemar (1988): Culture Barriers as a Social Psychological Construct: An Empirical Validation. In: Young Y. Kim (Eds.): Cross cultural adaptation. Current approaches. Newberry Park: SAGE, pp. 168–190.
Torikai, Kumiko (2009): Voices of the invisible presence: diplomatic interpreters in post-World War II Japan. Amsterdam: John Benjamins Pub.
Tubilewicz, Czeslaw (2006): Critical issues in contemporary China. London: Routledge.
Turner, Graeme; Tay, Jinna (Eds.) (2009): Television studies after TV: understanding television in the post- broadcast era. Oxen: Routledge.
Vaidya, Ashish K. (2006): Globalization: encyclopedia of trade, labor, and politics. Santa Barbara: ABC-CLIO.
Varela, Frederic C. (1997): Translating non-verbal information in dubbing. In: Fernando Poyatos (Eds.): Nonverbal communication and translation: new perspectives and challenges in literature, interpretation and the media. Amsterdam: J. Benjamins, pp. 315–326.
Verderber, Kathleen S.; Verderber, Rudolph F.; Sellnow, Deanna D. (2010): Communicate! Boston: Cengage Learning.
Vogel, Harold L. (2001): Entertainment industry economics. A guide for financial analysis. Cambridge: Cambridge Univ. Press.
Walker, Allan; Dimmock, Clive A. (Eds.) (2002): School leadership and administration: adopting a cultural perspective. New York: RoutledgeFalmer.
Wang, Yuan-Kang (2010): Harmony and war: Confucian culture and Chinese power politics. New York: Columbia University Press.
Ward, C.A; Bochner, S.; Furnham, A. (2001): The psychology of culture shock. London: Routledge.
Ward, Colleen; Kennedy, Anthony (1999): The measurement of socialcultural adaptation. In: *International Journal of Intercultural Relations* (23), pp. 659–677.
Watkins, David (2002): Learning and teaching. A cross-cultural perspective. In: Allan Walker and Clive A. Dimmock (Eds.): School leadership and ad-

ministration: adopting a cultural perspective. New York: Routledge Falmer, pp. 61–76.

Watkins, Ryan; Leigh, Doug (2009): Handbook of Improving Performance in the Workplace. Volume 2: Selecting and Implementing Performance Interventions. New York: John Wiley & Sons.

Wemheuer, Felix; Manning, Kimberley E. (Eds.) (2011): Eating bitterness: new perspectives on China's Great Leap Forward and famine. Vancouver: UBC Press.

Wen, Julie J.; Tisdell, Clement A. (2001): Tourism and China's development: policies, regional economic growth and ecotourism. Singapore: World Scientific.

Wijgers, Alexander (2008): Geography and regional development in China. Norderstedt: GRIN Verlag GmbH.

Wolther, Irving (2006): "Kampf der Kulturen": der Eurovision Song Contest als Mittel national-kultureller Repräsentation. Würzburg: Königshausen und Neumann.

Wong, Lornita Y.F. (1992): Education of Chinese children in Britain and the USA. Clevedon: Multilingual matters.

Wood, J.T (2010): Communication Mosaics: An Introduction to the Field of Communication. Boston: Cengage Learning.

Wood, Julia T. (2009): Interpersonal Communication: Everyday Encounters. Boston: Cengage Learning.

Wu, Xinxun; Jin, Guanjun (2004): China media economics research 1949 - 2004 (中国传媒经济研究 1949 - 2004). Shanghai: Fudan University.

Wu, Xinxun; Jin, Guanjun (Eds.) (2004): Media economics research in east and west China (中国东西部传媒经济研究). Shanghai: Shanghai University.

Xi, Jieying; Sun, Yunxiao; Xiao, Jingjian (Eds.) (2006): Chinese youth in transition. Hampshire: Ashgate.

Xu, Luo (2002): Searching for life's meaning: changes and tensions in the worldviews of Chinese youth in the 1980s. Michigan: University of Michigan Press.

Yang, Fenggang (2012): Religion in China. New York: Oxford University Press.

Yeshin, Tony (1998): Integrated marketing communications. The holistic approach. Oxford: Butterworth Heinemann.

Zandpour, Fred; Sadri, Golnaz (1996): Communication in personal relationships in Iran: A comparative analysis. In: W.B Gudykunst, S. Ting-Toomey and T. Nishida (Eds.): Communication in personal relationships across cultures. Thousand Oaks: Sage Publications, pp. 174–196.

Zeng, Douglas Z. (2010): Building engines for growth and competitiveness in China: experience with special economic zones and industrial clusters. Washington DC: World Bank.

Zhang, Xiaoling (2009): From "foreign propaganda" to "international communication": China's promotion of soft power in the age of information and communication technologies. In: Xiaoling Zhang and Yongnian Zheng (Eds.): China's information and communications technology revolution: social changes and state responses. Oxon: Routledge, pp. 103–120.

Zhang, Xiaoling; Zheng, Yongnian (Eds.) (2009): China's information and communications technology revolution: social changes and state responses. Oxon: Routledge.

Zhao, Suisheng (2000): China and democracy: the prospect for a democratic China. New York: Routledge.

Zhao, Xia (2006): Chinese childrens consumption. In: Jieying Xi, Yunxiao Sun and Jingjian Xiao (Eds.): Chinese youth in transition. Hampshire: Ashgate, pp. 59–69.

Zhao, Yuezhi (1998): Media, market, and democracy in China. Between the party line and the bottom line. Urbana: Univ. of Illinois Press.

Zhou, Yanyu (2008): Success for all: a comprehensive educational reform for improving at-risk students in an urban school in China. Charlotte: IAP.

Zhu, Ying; Keane, Michael; Bai, Ruoyun (Eds.) (2008): TV drama in China. Hong Kong: Hong Kong University Press.

Zou, Xiaoli; Wang, Yonglian (2003): The market share competition of satellite television (卫视的市场份额竞争). Anhui: Anhui Television Station.

# Appendix A: Television Schedule Analysis

## Appendix 1: Code Book

| Part A | Technical Variables |
|---|---|
| Variable 1 | Number |
| Variable 2 | Name |
| Variable 3 | City |
| Variable 4 | TV channel |
| Variable 5 | Start time (dd-mm-yyyy hh:mm) |
| Variable 6 | End time (dd-mm-yyyy hh:mm) |
| Variable 7 | Duration (in hour) |
| **Part B** | **Cultural Difference Variables** |
| B1: Content Catalogue | |
| Variable 8 | Genre |
| Variable 9 | Theme |
| B2: Cultural Modification | |
| Variable 10 | Modification type I: Content modification |
| Variable 11 | Modification type II: Language modification |
| Variable 12 | Modification type III: Person modification |
| B3: Cultural Proximity | |
| Variable 13 | Country of origin |
| Variable 14 | Cultural proximity |

### Part A: Technical Variables

Variable 1: Number
For example: 101008

Variable 2: Name
For example: *Animals World*

Variable 3: City

| Code | Description |
|---|---|
| 1 | Beijing |
| 2 | Shanghai |
| 3 | Guangzhou |
| 4 | Nanjing |
| 5 | Xi'an |

Variable 4: TV Channel
For example: *Chinese Central Television* Channel 1 is "CCTV-1"

Variable 5: Start Time (dd-mm-yyyy hh:mm)

Variable 6: End Time (dd-mm-yyyy hh:mm)

Variable 7: Duration (in hour)

**Teil B: Cultural Difference Variables**

**B1: Content Catalogue**

Variable 8: Genre

| Code | Description |
|---|---|
| 0 | Other |
| 1 | Movie |
| 2 | Television series |
| 3 | Sports program |
| 4 | Documentary |
| 5 | Music program |
| 6 | Entertainment show |
| 7 | News program |

Variable 9: Theme

| Code | Description | Code | Description |
|---|---|---|---|
| 1 | *Movie* | 314 | Skateboard |
| 100 | Other movie | 315 | Skating |
| 101 | Action | 316 | Skiing |
| 102 | Adventure | 317 | Snooker |
| 103 | Biography | 318 | Strong man |
| 104 | Cartoon | 319 | Swimming |
| 105 | Children's | 320 | Tennis |
| 106 | Crime | 321 | Track & field |
| 107 | Family | 322 | Water ball |
| 108 | Friendship | 323 | Wrestling |
| 109 | Romance | 324 | Mixed (magazine) |
| 110 | Science fiction | 4 | *Documentary* |
| 111 | War | 401 | Accidents/disasters |
| 112 | Mixed (magazine) | 402 | Biography |
| 2 | *TV series* | 403 | Crime |
| 200 | Other TV series | 404 | Geography |
| 201 | Action | 405 | Magic |
| 202 | Cartoon | 406 | Nature/animal/plant |
| 203 | Children's | 407 | Science/technology |
| 204 | Comedy | 408 | War |
| 205 | Romance | 409 | Mixed (magazine) |
| 3 | *Sports program* | 5 | *Music program* |
| 300 | Other sports program | 500 | Other music program |
| 301 | Air race | 501 | Classical music |
| 302 | Badminton | 502 | Film music |
| 303 | Baseball | 503 | Music story |
| 304 | Basketball | 504 | Music television (MTV) |

| | | | |
|---|---|---|---|
| 305 | Boxing | 6 | *Entertainment show* |
| 306 | Curling | 600 | Other entertainment show |
| 307 | Bullfighting | 601 | Game show |
| 308 | Football | 602 | Magic/acrobatic |
| 309 | Golf | 7 | *News program* |
| 310 | Gymnastics | 700 | Other news program |
| 311 | Motor racing | 701 | News report |
| 312 | National football | 702 | News magazine |
| 313 | Ocean race | | |

## B2: Cultural Modification

### Variable 10: Modification Type I: Content Modification

| Code | Description |
|---|---|
| 1 | Foreign content without content modification |
| 2 | Foreign content with some content modification |
| 3 | Foreign content with some added local content |
| 4 | Local content with a foreign format |

### Variable 11: Modification Type II: Language Modification

| Code | Description |
|---|---|
| 1 | Original foreign language without subtitles |
| 2 | Original foreign language with subtitles |
| 3 | Dubbing |
| 4 | Replacement of original spoken dialogue |

### Variable 12: Modification Type III: Person Modification (Actor/Presenter/Participant)

| Code | Description |
|---|---|
| 1 | No person replacement |
| 2 | Some person replacement |

| | |
|---|---|
| 3 | Complete person replacement |

## B3: Cultural Proximity

Variable 13: Country of Origin

| Code | Description | Code | Description |
|---|---|---|---|
| 0 | Classification not possible | 30 | Countries in Asian |
| 10 | Countries in North America | 31 | Japan |
| 11 | United States | 32 | Korean |
| 12 | Canada | 33 | Thailand |
| 13 | Other North America countries | 34 | Vietnam |
| 20 | Countries in Europe | 35 | Other Asian countries |
| 21 | United Kingdom | 40 | Countries in Oceania |
| 22 | Germany | 50 | Countries in South American |
| 23 | France | 60 | Countries in Africa |
| 24 | Spain | | |
| 25 | Italy | | |
| 26 | Other European countries | | |

Variable 14: Degree of Cultural Proximity

| Code | Description | Sample Countries |
|---|---|---|
| 0 | Classification not possible | |
| 1 | Very similar | Asian countries: Japan, Korean |
| 2 | Similar | Asian countries: Thailand, Vietnam, others |
| 3 | Different | North American countries: U.S., Canada<br>European countries: UK, Germany, France, Italy<br>Other countries: Australia, New Zealand |
| 4 | Very different | South American countries: Brazil, Argentina<br>African countries: South Africa |

# Appendix B: Interview

## Appendix 2: List of Interviewees

| Nr. | Name | Gender | Age | Education | Occupation | Viewer Type |
|---|---|---|---|---|---|---|
| BJ_1 | Wang | m | 31 | Bachelor | | B: Learner |
| BJ_2 | Ding | w | 28 | Bachelor | | D: Adjuster |
| BJ_3 | Li | m | 38 | Middle School | | A: Escapee |
| BJ_4 | Peng | w | 25 | Bachelor | | E: Neutralist |
| BJ_5 | Yu | m | 43 | Master | | E: Neutralist |
| BJ_6 | Feng | w | 27 | Bachelor | Student | D: Adjuster |
| BJ_7 | Wu | m | 34 | Master | Manager | D: Adjuster |
| SH_1 | Huang | m | 31 | Doctor | College Teacher | D: Adjuster |
| SH_2 | Chen | w | 43 | Middle School | | A: Escapee |
| SH_3 | Wang | w | 36 | Master | Advertising Agent | B: Learner |
| SH_4 | Zhang | m | 22 | Bachelor | Student | B: Learner |
| SH_5 | Fang | m | 23 | Bachelor | Student | B: Learner |
| GZ_1 | Cai | m | 39 | Master | Media Designer | D: Adjuster |
| GZ_2 | Yang | w | 41 | Bachelor | International Trader | D: Adjuster |
| GZ_3 | Zhang | m | 27 | Master | International Trader | D: Adjuster |
| GZ_4 | Guo | w | 25 | Bachelor | | A: Escapee |
| GZ_5 | Pan | m | 28 | Master | | B: Learner |
| GZ_6 | Shi | w | 35 | Master | Lawyer | D: Adjuster |
| NJ_1 | Yuan | m | 28 | Bachelor | | A: Escapee |
| NJ_2 | Fan | w | 36 | Bachelor | Mother | A: Escapee |
| NJ_3 | Hu | m | 34 | Bachelor | | A: Escapee |
| NJ_4 | Lin | m | 23 | Bachelor | | -------------- |
| NJ_5 | Cui | m | 27 | Bachelor | | D: Adjuster |
| NJ_6 | Zheng | w | 17 | Middle School | Student | B: Learner |
| NJ_7 | Jiang | m | 41 | Bachelor | Father | A: Escapee |

| | | | | | | |
|---|---|---|---|---|---|---|
| NJ_8 | Zhao | w | 43 | Middle School | Worker | A: Escapee |
| XA_1 | Yin | m | 19 | Bachelor | | A: Escapee |
| XA_2 | Song | m | 26 | Bachelor | | B: Learner |
| XA_3 | Han | m | 27 | Bachelor | | D: Adjuster |
| XA_4 | Meng | w | 25 | Bachelor | Student | B: Learner |
| XA_5 | Feng | w | 12 | Primary school | | B: Learner |
| XA_6 | Ma | m | 37 | Middle school | | -------------- |
| XA_7 | Zhu | w | 31 | Doctor | Ph.D. Student | B: Learner |
| XA_8 | Xu | m | 33 | Master | | B: Learner |
| XA_9 | Zhou | w | 34 | Bachelor | | B: Learner |
| XA_10 | Zhang | m | 29 | Master | Scholar | B: Learner |

Note: BJ = Beijing, GZ = Guangzhou, NJ = Nanjing, SH = Shanghai, XA = Xi'An

# Appendix 3: Usage of Foreign Television Programs

| | Usage | Movie | TV se. | Sport | Docu. | Enterta. | News | Music |
|---|---|---|---|---|---|---|---|---|
| BJ_1 | Medium | x | x | x | | | | |
| BJ_2 | Much | x | x | x | | | | x |
| BJ_3 | None | | | | | | | |
| BJ_4 | Medium | | | | | | | |
| BJ_5 | Much | x | x | | | | | |
| BJ_6 | Little | x | x | | | | | |
| BJ_7 | Much | x | x | | x | x | x | x |
| SH_1 | Little | x | x | x | x | x | | x |
| SH_2 | Little | x | | | | | | |
| SH_3 | Much | x | x | x | x | | | x |
| SH_4 | Medium | x | x | | x | | | |
| SH_5 | Much | x | x | x | x | x | | |
| GZ_1 | Medium | x | x | x | x | x | x | x |
| GZ_2 | Much | x | x | x | x | x | x | x |
| GZ_3 | Much | | x | x | x | x | | x |
| GZ_4 | Medium | x | x | x | | | x | |
| GZ_5 | Little | x | | | | | | |
| GZ_6 | Medium | x | | | x | | | |
| NJ_1 | Little | x | x | | | | | |
| NJ_2 | Little | x | x | | x | | | |
| NJ_3 | Medium | x | x | | | | | |
| NJ_4 | Medium | | | | x | | x | |
| NJ_5 | Little | | x | x | | | | |
| NJ_6 | Little | x | x | | x | | | x |
| NJ_7 | Medium | | x | | | | | |
| NJ_8 | Little | | | x | | | x | |
| XA_1 | Little | x | | | | | | |

| | | | | | | | | |
|---|---|---|---|---|---|---|---|---|
| XA_2 | Much | x | x | x | | | | |
| XA_3 | Medium | x | x | | | | | |
| XA_4 | Little | | | | x | | | |
| XA_5 | Little | | x | | | | | |
| XA_6 | Little | | | x | | | | |
| XA_7 | Little | | | | | | | |
| XA_8 | Medium | x | x | | | | | |
| XA_9 | Little | x | | | | | | |
| XA_10 | Much | x | x | x | x | | | |

Note: TV se. = Television series, Sport = sports program, Docu. = documentary, Enterta. = entertainment program, News = news program, Music = music programs.

# Appendix C: Survey

## Appendix 4: Survey (English)

Survey No.:

### A Survey about Cultural Differences in Foreign Television Programs in China

**"*Cultural difference*" means difference between background culture in foreign television programs and Chinese culture.**

"*Foreign television programs*" include foreign TV programs broadcast by Chinese television stations and downloaded from the internet or watched online.

"*Foreign television programs*" are foreign movies, television series, sports programs, documentaries, entertainment shows, news programs, and so on.

"*Foreign country*" includes all countries in the world (except China), such as the United States, European countries, Japan, and Korea. Television programs from Hong Kong, Taiwan and Macau are not classed as foreign programs.

Please evaluate particular cultural differences one after another and tick the boxes.

(a) First please evaluate how large the particular cultural differences are. "**1**" means this cultural difference is **very small**. "**5**" means this cultural difference is **very large**. Please see the table below:

| Very small | Small | Normal | Large | Very Large |
|---|---|---|---|---|
| 1 ------------- | 2 ------------- | 3 ------------- | 4 ------------- | 5 |

(b) Please evaluate to what extent you accept the particular cultural differences. "**1**" means you can **accept** this cultural difference **totally**. "**5**" means you **cannot accept** this cultural difference. Please see this table:

| Total | Most | Normal | Little | Not |
|---|---|---|---|---|
| 1 ------------- | 2 ------------- | 3 ------------- | 4 ------------- | 5 |

## 1. Movie and Television Series

1.1. How many foreign movies and television series do you watch, e.g., "Titanic", "Desperate Housewives", and Disney cartoons?

☐ Very much ☐ Much ☐ Normal ☐ Little ☐ Very little

1.2. To what extent could you perceive the following cultural differences in foreign movies and TV series? To what extent could you accept them?

| | **Perception**<br>very small → very large<br>1 — 2 — 3 — 4 — 5 | **Acceptance**<br>totally can → cannot<br>1 — 2 — 3 — 4 — 5 |
|---|---|---|
| Appearance of character (e.g., physical appearance, costume, make-up) | ☐ ☐ ☐ ☐ ☐ | ☐ ☐ ☐ ☐ ☐ |
| Personality of character (e.g., facial expression, gesture) | ☐ ☐ ☐ ☐ ☐ | ☐ ☐ ☐ ☐ ☐ |
| Music style (e.g., background music, title song) | ☐ ☐ ☐ ☐ ☐ | ☐ ☐ ☐ ☐ ☐ |
| Theme of drama plot | ☐ ☐ ☐ ☐ ☐ | ☐ ☐ ☐ ☐ ☐ |
| Proximity to daily life of drama plot | ☐ ☐ ☐ ☐ ☐ | ☐ ☐ ☐ ☐ ☐ |
| Fantasy of drama plot | ☐ ☐ ☐ ☐ ☐ | ☐ ☐ ☐ ☐ ☐ |

1.3. How much do you like foreign movies and TV series from these countries (except China)?

| | |
|---|---|
| Asia (e.g., Japan, Korea, India) | ☐ very like ☐ like ☐ normal ☐ little like ☐ not like |
| America (e.g., USA, Canada) | ☐ very like ☐ like ☐ normal ☐ little like ☐ not like |
| Europe (e.g., UK, Germany, Italy) | ☐ very like ☐ like ☐ normal ☐ little like ☐ not like |
| Other Country | ☐ very like ☐ like ☐ normal ☐ little like ☐ not like |

1.4. Which foreign movies and TV series (incl. cartoons) do you **like** to watch? Please give some examples.

1.5. Which foreign movies and TV series (incl. cartoons) do you **not like** to watch? Please give some examples.

——

**2. Sports Program**

2.1. How many foreign sports programs do you watch, e.g., "NBA", "Premiership", "Serie A", and "Marlboro World of Sports"?

☐ Very much ☐ Much ☐ Normal ☐ Little ☐ Very little

2.2. To what extent could you perceive the following cultural differences in foreign sports programs? To what extent could you accept them?

| | **Perception**<br>very small → very large<br>1 — 2 — 3 — 4 — 5 | **Acceptance**<br>totally can → cannot<br>1 — 2 — 3 — 4 — 5 |
|---|---|---|
| Moderation/commentary | ☐ ☐ ☐ ☐ ☐ | ☐ ☐ ☐ ☐ ☐ |
| Sports culture | ☐ ☐ ☐ ☐ ☐ | ☐ ☐ ☐ ☐ ☐ |
| Commercialization | ☐ ☐ ☐ ☐ ☐ | ☐ ☐ ☐ ☐ ☐ |
| Sports types (e.g., basketball, football, boxing, wrestling, fomular 1) | ☐ ☐ ☐ ☐ ☐ | ☐ ☐ ☐ ☐ ☐ |

2.3. How much do you like foreign sport programs from these countries (except China)?

| | |
|---|---|
| Asia (e.g., Japan, Korea, India) | ☐ very like ☐ like ☐ normal ☐ little like ☐ not like |
| America (e.g., USA, Canada) | ☐ very like ☐ like ☐ normal ☐ little like ☐ not like |
| Europe (e.g., UK, Germany, Italy) | ☐ very like ☐ like ☐ normal ☐ little like ☐ not like |
| Other Country | ☐ very like ☐ like ☐ normal ☐ little like ☐ not like |

2.4. Which foreign sport programs do you **like** to watch? Please give some examples.

——

2.5. Which foreign sport programs do you **not like** to watch? Please give some examples.

____

**3. Documentary**

3.1. How many foreign documentaries do you watch, e.g., "Discovery" and "GEO"?

☐ Very much ☐ Much ☐ Normal ☐ Little ☐ Very little

3.2. To what extent could you perceive the following cultural differences in foreign documentaries? To what extent could you accept them?

| | **Perception**<br>very small → very large<br>1 — 2 — 3 — 4 — 5 | **Acceptance**<br>totally can → cannot<br>1 — 2 — 3 — 4 — 5 |
|---|---|---|
| Objectivity vs. subjectivity | ☐ ☐ ☐ ☐ ☐ | ☐ ☐ ☐ ☐ ☐ |
| Explicit vs. implicit | ☐ ☐ ☐ ☐ ☐ | ☐ ☐ ☐ ☐ ☐ |
| Starting point/perspective | ☐ ☐ ☐ ☐ ☐ | ☐ ☐ ☐ ☐ ☐ |
| Theme (e.g., nature, geographic) | ☐ ☐ ☐ ☐ ☐ | ☐ ☐ ☐ ☐ ☐ |

3.3. How much do you like foreign documentaries from these countries (except China)?

| | |
|---|---|
| Asia (e.g., Japan, Korea, India) | ☐ very like ☐ like ☐ normal ☐ little like ☐ not like |
| America (e.g., USA, Canada) | ☐ very like ☐ like ☐ normal ☐ little like ☐ not like |
| Europe (e.g., UK, Germany, Italy) | ☐ very like ☐ like ☐ normal ☐ little like ☐ not like |
| Other Country | ☐ very like ☐ like ☐ normal ☐ little like ☐ not like |

3.4. Which foreign documentaries do you **like** to watch? Please give some examples.

____

3.5. Which foreign documentaries do you **not like** to watch? Please give some examples.

——

**4. Entertainment Show**

4.1. How many foreign entertainment shows do you watch, e.g., "Survivor" and "Chao Ji Bian Bian Bian"?

☐ Very much ☐ Much ☐ Normal ☐ Little ☐ Very little

4.2. To what extent could you perceive the following cultural differences in foreign entertainment shows? To what extent could you accept them?

| | **Perception**<br>very small → very large<br>1 — 2 — 3 — 4 — 5 | **Acceptance**<br>totally can → cannot<br>1 — 2 — 3 — 4 — 5 |
|---|---|---|
| Speaking style of presenter | ☐ ☐ ☐ ☐ ☐ | ☐ ☐ ☐ ☐ ☐ |
| Content of entertainment (e.g., game, talent, dating, quiz show) | ☐ ☐ ☐ ☐ ☐ | ☐ ☐ ☐ ☐ ☐ |

4.3. How much do you like foreign entertainment shows from these countries (except China)?

| | |
|---|---|
| Asia (e.g., Japan, Korea, India) | ☐ very like ☐ like ☐ normal ☐ little like ☐ not like |
| America (e.g., USA, Canada) | ☐ very like ☐ like ☐ normal ☐ little like ☐ not like |
| Europe (e.g., UK, Germany, Italy) | ☐ very like ☐ like ☐ normal ☐ little like ☐ not like |
| Other Country | ☐ very like ☐ like ☐ normal ☐ little like ☐ not like |

4.4. Which foreign entertainment shows do you **like** to watch? Please give some examples.

——

4.5. Which foreign entertainment shows do you **not like** to watch? Please give some examples.

——

## 5. News Program

5.1. How many foreign news programs do you watch, e.g., "CNN news" and "BBC news"?

☐ Very much ☐ Much ☐ Normal ☐ Little ☐ Very little

5.2. To what extent could you perceive the following cultural differences in foreign news programs? To what extent could you accept them?

| | **Perception**<br>very small → very large<br>1 — 2 — 3 — 4 — 5 | **Acceptance**<br>totally can → cannot<br>1 — 2 — 3 — 4 — 5 |
|---|---|---|
| Appearance of moderator (e.g., physical appearance, costume, make-up) | ☐ ☐ ☐ ☐ ☐ | ☐ ☐ ☐ ☐ ☐ |
| Personality of moderator (e.g., facial expression, gesture) | ☐ ☐ ☐ ☐ ☐ | ☐ ☐ ☐ ☐ ☐ |
| Objectivity vs. subjectivity of commentary | ☐ ☐ ☐ ☐ ☐ | ☐ ☐ ☐ ☐ ☐ |
| Starting point/perspective of commentary | ☐ ☐ ☐ ☐ ☐ | ☐ ☐ ☐ ☐ ☐ |
| Diversity of news theme | ☐ ☐ ☐ ☐ ☐ | ☐ ☐ ☐ ☐ ☐ |
| Sensitivity of news theme | ☐ ☐ ☐ ☐ ☐ | ☐ ☐ ☐ ☐ ☐ |

5.3. How much do you like foreign news programs from these countries (except China)?

| | |
|---|---|
| Asia (e.g., Japan, Korea, India) | ☐ very like ☐ like ☐ normal ☐ little like ☐ not like |
| America (e.g., USA, Canada) | ☐ very like ☐ like ☐ normal ☐ little like ☐ not like |
| Europe (e.g., UK, Germany, Italy) | ☐ very like ☐ like ☐ normal ☐ little like ☐ not like |
| Other Country | ☐ very like ☐ like ☐ normal ☐ little like ☐ not like |

5.4. Which foreign news programs do you **like** to watch? Please give some examples.

5.5. Which foreign news programs do you **not like** to watch? Please give some examples.

——

## 6. Music Program

6.1. How many foreign music programs do you watch, e.g., "MTV" and "Vienna New Year's Concert"?

☐ Very much ☐ Much ☐ Normal ☐ Little ☐ Very little

6.2. To what extent could you perceive the following cultural differences in foreign music programs? To what extent could you accept them?

| | **Perception**<br>very small → very large<br>1 — 2 — 3 — 4 — 5 | **Acceptance**<br>totally can → cannot<br>1 — 2 — 3 — 4 — 5 |
|---|---|---|
| Music style | ☐ ☐ ☐ ☐ ☐ | ☐ ☐ ☐ ☐ ☐ |
| Music quality | ☐ ☐ ☐ ☐ ☐ | ☐ ☐ ☐ ☐ ☐ |

6.3. How much do you like foreign music programs from these countries (except China)?

| | |
|---|---|
| Asia (e.g., Japan, Korea, India) | ☐ very like ☐ like ☐ normal ☐ little like ☐ not like |
| America (e.g., USA, Canada) | ☐ very like ☐ like ☐ normal ☐ little like ☐ not like |
| Europe (e.g., UK, Germany, Italy) | ☐ very like ☐ like ☐ normal ☐ little like ☐ not like |
| Other Country | ☐ very like ☐ like ☐ normal ☐ little like ☐ not like |

6.4. Which foreign music programs do you **like** to watch? Please give some examples.

——

6.5. Which foreign music programs do you **not like** to watch? Please give some examples.

——

..........................................................................................................................................

## 7. Language

7.1. Which form of language translation in foreign television programs would you rather like?

☐ Dubbing

☐ Original language with subtitles

☐ Original language without subtitles

☐ I don´t care

7.2. To what extent could you perceive the following language differences in foreign television programs? To what extent could you accept them?

| | **Perception**<br>very small → very large<br>1 — 2 — 3 — 4 — 5 | **Acceptance**<br>totally can → cannot<br>1 — 2 — 3 — 4 — 5 |
|---|---|---|
| Voice of dubbing | ☐ ☐ ☐ ☐ ☐ | ☐ ☐ ☐ ☐ ☐ |
| Emotion of dubbing | ☐ ☐ ☐ ☐ ☐ | ☐ ☐ ☐ ☐ ☐ |
| Coordination of dubbing | ☐ ☐ ☐ ☐ ☐ | ☐ ☐ ☐ ☐ ☐ |
| Accuracy of translation | ☐ ☐ ☐ ☐ ☐ | ☐ ☐ ☐ ☐ ☐ |
| Humor in translation | ☐ ☐ ☐ ☐ ☐ | ☐ ☐ ☐ ☐ ☐ |

## 8. Value

8.1. To what extent could you perceive the following cultural differences of values in foreign television programs? To what extent could you accept them?

| | **Perception**<br>very small → very large<br>1 — 2 — 3 — 4 — 5 | **Acceptance**<br>totally can → cannot<br>1 — 2 — 3 — 4 — 5 |
|---|---|---|
| Individualism | ☐ ☐ ☐ ☐ ☐ | ☐ ☐ ☐ ☐ ☐ |
| Religion | ☐ ☐ ☐ ☐ ☐ | ☐ ☐ ☐ ☐ ☐ |
| Morality (e.g., moral standard) | ☐ ☐ ☐ ☐ ☐ | ☐ ☐ ☐ ☐ ☐ |
| Family (e.g., family-centered) | ☐ ☐ ☐ ☐ ☐ | ☐ ☐ ☐ ☐ ☐ |
| Liberty/democracy/equality | ☐ ☐ ☐ ☐ ☐ | ☐ ☐ ☐ ☐ ☐ |

| | | |
|---|---|---|
| Social status/competition | ☐ ☐ ☐ ☐ ☐ | ☐ ☐ ☐ ☐ ☐ |
| Patriotism | ☐ ☐ ☐ ☐ ☐ | ☐ ☐ ☐ ☐ ☐ |

**9. Way of Life**

9.1. To what extent could you perceive the following cultural differences of way of life in foreign television programs? To what extent could you accept them?

| | **Perception**<br>very small → very large<br>1 — 2 — 3 — 4 — 5 | **Acceptance**<br>totally can → cannot<br>1 — 2 — 3 — 4 — 5 |
|---|---|---|
| Habits/customs/rituals | ☐ ☐ ☐ ☐ ☐ | ☐ ☐ ☐ ☐ ☐ |
| Living environment | ☐ ☐ ☐ ☐ ☐ | ☐ ☐ ☐ ☐ ☐ |

## 10. Education Method

10.1. To what extent could you perceive the following cultural differences of education method in foreign television programs? To what extent could you accept them?

| | **Perception**<br>very small → very large<br>1 — 2 — 3 — 4 — 5 | **Acceptance**<br>totally can → cannot<br>1 — 2 — 3 — 4 — 5 |
|---|---|---|
| Freedom/independence | ☐ ☐ ☐ ☐ ☐ | ☐ ☐ ☐ ☐ ☐ |
| Imagination/creativity | ☐ ☐ ☐ ☐ ☐ | ☐ ☐ ☐ ☐ ☐ |
| Equality between young and old | ☐ ☐ ☐ ☐ ☐ | ☐ ☐ ☐ ☐ ☐ |
| Demand on children | ☐ ☐ ☐ ☐ ☐ | ☐ ☐ ☐ ☐ ☐ |

..........................................................................................................

## 11. Summary

11.1. How large are cultural differences overall?

☐ Very small ☐ Small ☐ Normal ☐ Large ☐ Very large

11.2. To what extent could you accept such cultural differences overall?

☐ Total ☐ Most ☐ Normal ☐ Little ☐ Not

## 12. Personal Information

12.1. What's your gender?

☐ Male ☐ Female

12.2. How old are you?

☐ 20 - 29 ☐ 30 - 39 ☐ 40 – 49

12.3. In which city are you living now?

☐ Beijing ☐ Shanghai ☐ Guangzhou ☐ Nanjing ☐ Xi'An

12.4. What is your highest level of education?

☐ Primary/middle school ☐ Bachelor ☐ Master ☐ Doctorate

12.5. How much are your income monthly?

☐ ¥ 0-999 ☐ ¥ 1 000-4 999 ☐ ¥ 5 000-9 999 ☐ over ¥ 10 000

12.6. How much do you know about foreign cultures?

☐ Very much ☐ Much ☐ Normal ☐ Little ☐ Very little

| **Please fill in your contact information (this is voluntary)** | | | |
|---|---|---|---|
| Name | | QQ | |
| Email | | MSN | |
| Telephone | | Skype | |
| Address | | | |

**You have successfully finished this questionnaire!**

**Thank you very much for your participation!**

# Appendix 5: Survey (Chinese)

问卷号:　　　（由工作人员填写）

## 关于外国电视节目中文化差异的问卷调查

欢迎您参加此次问卷调查!

**“文化差异”是指外国电视节目中所反映的文化与中国文化之间的差异。**

**“外国电视节目”** 包括所有由中国电视台播放的外国电视节目，以及从互联网上下载的或者是在线收看的外国电视节目。

**“外国电视节目的类型”** 比如：外国电影，电视剧，体育节目，纪录片，娱乐节目，新闻节目等等。

**“外国”** 在这里是指除中国以外的所有国家，比如美国，欧洲国家，日本，韩国等等。港澳台的电视节目不属于外国节目。

**填写说明：** （本页只是示例，正式问题从下一页开始）

请您在此问卷中，对各种不同类型的文化差异逐一做出判断，并且在相应的方框里打钩。示例：☑

（1）首先请判断，外国电视节目中所反映的文化与中国文化相比，他们之间的差异有多大, 详见下图:

- 数字"**1**”表示"文化差异**非常小**”

- 数字"**5**”表示"文化差异**非常大**”

| 非常小 | 比较小 | 一般 | 比较大 | 非常大 |
|---|---|---|---|---|
| 1 | 2 | 3 | 4 | 5 |

（2）接着请判断，您在何种程度上可以接受外国电视节目中存在的这些文化差异, 详见下图:

- 数字"**1**”表示"**完全能**接受"

- 数字"**5**”表示"**完全不能**接受"

| 完全能 | 大部分能 | 一半一半 | 小部分能 | 完全不能 |
|---|---|---|---|---|
| 1 | 2 | 3 | 4 | 5 |

## 1. 电影和电视剧

1.1. 您看过多少国外的电影和电视剧，包括动画片？比如《泰坦尼克号》，《疯狂主妇》，迪斯尼动画片。

☐ 非常多 ☐ 比较多 ☐ 一般 ☐ 比较少 ☐ 非常少

1.2. 外国电影电视剧中所反映的文化与中国文化相比，他们之间的文化差异有多大？您在何种程度上可以接受这些文化差异？

| | **差异大小**<br>非常小 → 非常大<br>1 — 2 — 3 — 4 — 5 | **接受程度**<br>完全能 → 完全不能<br>1 — 2 — 3 — 4 — 5 |
|---|---|---|
| 人物的 外表（比如外貌，着装，化妆） | ☐ ☐ ☐ ☐ ☐ | ☐ ☐ ☐ ☐ ☐ |
| 人物的 个性（比如表情，动作） | ☐ ☐ ☐ ☐ ☐ | ☐ ☐ ☐ ☐ ☐ |
| 音乐风格（比如背景音乐，主题曲） | ☐ ☐ ☐ ☐ ☐ | ☐ ☐ ☐ ☐ ☐ |
| 剧情的 题材 | ☐ ☐ ☐ ☐ ☐ | ☐ ☐ ☐ ☐ ☐ |
| 剧情的 生活贴近性 | ☐ ☐ ☐ ☐ ☐ | ☐ ☐ ☐ ☐ ☐ |
| 剧情的 想象力 | ☐ ☐ ☐ ☐ ☐ | ☐ ☐ ☐ ☐ ☐ |

1.3. 请您判断，您对以下国家或地区（除中国以外）的电影电视剧的喜爱程度。

| | |
|---|---|
| 亚洲（比如日本，韩国，印度） | ☐ 非常喜欢 ☐ 比较喜欢 ☐ 一般 ☐ 不太喜欢 ☐ 很不喜欢 |
| 美洲（比如美国，加拿大） | ☐ 非常喜欢 ☐ 比较喜欢 ☐ 一般 ☐ 不太喜欢 ☐ 很不喜欢 |
| 欧洲（比如英国，德国，意大利） | ☐ 非常喜欢 ☐ 比较喜欢 ☐ 一般 ☐ 不太喜欢 ☐ 很不喜欢 |
| 其他国家 | ☐ 非常喜欢 ☐ 比较喜欢 ☐ 一般 ☐ 不太喜欢 ☐ 很不喜欢 |

1.4. 有哪些国外的电影电视剧（包括动画片）您**喜欢**看？请举例。

——

1.5. 有哪些国外的电影电视剧（包括动画片）您**不喜欢**看？请举例。

——

**2. 体育节目**

2.1. 您看过多少国外的体育节目？比如：NBA，意甲，英超，《万宝路体育世界》。

☐ 非常多 ☐ 比较多 ☐ 一般 ☐ 比较少 ☐ 非常少

2.2. 外国体育节目中所反映的文化与中国文化相比，他们之间的文化差异有多大？您在何种程度上可以接受这些文化差异？

| | **差异大小**<br>非常小 → 非常大<br>1 — 2 — 3 — 4 — 5 | **接受程度**<br>完全能 → 完全不能<br>1 — 2 — 3 — 4 — 5 |
|---|---|---|
| 解说和主持的风格 | ☐ ☐ ☐ ☐ ☐ | ☐ ☐ ☐ ☐ ☐ |
| 体育文化 | ☐ ☐ ☐ ☐ ☐ | ☐ ☐ ☐ ☐ ☐ |
| 体育节目的商业化 | ☐ ☐ ☐ ☐ ☐ | ☐ ☐ ☐ ☐ ☐ |
| 体育类型（比如篮球，足球，拳击，摔跤，一级方程式赛车） | ☐ ☐ ☐ ☐ ☐ | ☐ ☐ ☐ ☐ ☐ |

2.3. 请您判断，您对以下国家或地区（除中国以外）的体育节目的喜爱程度。

| | |
|---|---|
| 亚洲（比如日本，韩国，印度） | ☐ 非常喜欢 ☐ 比较喜欢 ☐ 一般 ☐ 不太喜欢 ☐ 很不喜欢 |
| 美洲（比如美国，加拿大） | ☐ 非常喜欢 ☐ 比较喜欢 ☐ 一般 ☐ 不太喜欢 ☐ 很不喜欢 |
| 欧洲（比如英国，德国，意大利） | ☐ 非常喜欢 ☐ 比较喜欢 ☐ 一般 ☐ 不太喜欢 ☐ 很不喜欢 |
| 其他国家 | ☐ 非常喜欢 ☐ 比较喜欢 ☐ 一般 ☐ 不太喜欢 ☐ 很不喜欢 |

2.4. 有哪些国外的体育节目您**喜欢**看？请举例。

——

2.5. 有哪些国外的体育节目您**不喜欢**看？请举例。

——

---

**3. 纪录片**

3.1. 您看过多少国外的纪录片？比如：《探索发现》，《国家地理杂志》。

☐ 非常多　☐ 比较多　☐ 一般　☐ 比较少　☐ 非常少

3.2. 外国纪录片中所反映的文化与中国文化相比，他们之间的文化差异有多大？您在何种程度上可以接受这些文化差异？

| | **差异大小**<br>非常小 → 非常大<br>1 — 2 — 3 — 4 — 5 | **接受程度**<br>完全能 → 完全不能<br>1 — 2 — 3 — 4 — 5 |
|---|---|---|
| 主观性，客观性 | ☐ ☐ ☐ ☐ ☐ | ☐ ☐ ☐ ☐ ☐ |
| 阐述的明确性，委婉性 | ☐ ☐ ☐ ☐ ☐ | ☐ ☐ ☐ ☐ ☐ |
| 看待问题的出发点，角度 | ☐ ☐ ☐ ☐ ☐ | ☐ ☐ ☐ ☐ ☐ |
| 纪录片题材（比如自然，地理） | ☐ ☐ ☐ ☐ ☐ | ☐ ☐ ☐ ☐ ☐ |

3.3. 请您判断，您对以下国家或地区（除中国以外）的纪录片的喜爱程度。

| | |
|---|---|
| 亚洲（比如日本，韩国，印度） | ☐ 非常喜欢 ☐ 比较喜欢 ☐ 一般 ☐ 不太喜欢 ☐ 很不喜欢 |
| 美洲（比如美国，加拿大） | ☐ 非常喜欢 ☐ 比较喜欢 ☐ 一般 ☐ 不太喜欢 ☐ 很不喜欢 |
| 欧洲（比如英国，德国，意大利） | ☐ 非常喜欢 ☐ 比较喜欢 ☐ 一般 ☐ 不太喜欢 ☐ 很不喜欢 |
| 其他国家 | ☐ 非常喜欢 ☐ 比较喜欢 ☐ 一般 ☐ 不太喜欢 ☐ 很不喜欢 |

3.4. 有哪些国外的纪录片您**喜欢**看？请举例。

——

3.5. 有哪些国外的纪录片您**不喜欢**看？请举例。

——

---

**4. 娱乐节目**

4.1. 您看过多少国外的娱乐节目？比如：《幸存者》，《超级变变变》。

☐ 非常多　☐ 比较多　☐ 一般　☐ 比较少　☐ 非常少

4.2. 外国娱乐节目中所反映的文化与中国文化相比，他们之间的文化差异有多大？您在何种程度上可以接受这些文化差异？

| | **差异大小**<br>非常小 → 非常大<br>1 — 2 — 3 — 4 — 5 | **接受程度**<br>完全能 → 完全不能<br>1 — 2 — 3 — 4 — 5 |
|---|---|---|
| 主持风格 | ☐ ☐ ☐ ☐ ☐ | ☐ ☐ ☐ ☐ ☐ |
| 娱乐内容（比如游戏秀，达人秀，约会秀，竞猜秀） | ☐ ☐ ☐ ☐ ☐ | ☐ ☐ ☐ ☐ ☐ |

4.3. 请您判断，您对以下国家或地区（除中国以外）的娱乐节目的喜爱程度。

| | |
|---|---|
| 亚洲（比如日本，韩国，印度） | ☐ 非常喜欢 ☐ 比较喜欢 ☐ 一般 ☐ 不太喜欢 ☐ 很不喜欢 |
| 美洲（比如美国，加拿大） | ☐ 非常喜欢 ☐ 比较喜欢 ☐ 一般 ☐ 不太喜欢 ☐ 很不喜欢 |
| 欧洲（比如英国，德国，意大利） | ☐ 非常喜欢 ☐ 比较喜欢 ☐ 一般 ☐ 不太喜欢 ☐ 很不喜欢 |
| 其他国家 | ☐ 非常喜欢 ☐ 比较喜欢 ☐ 一般 ☐ 不太喜欢 ☐ 很不喜欢 |

4.4. 有哪些国外的娱乐节目您**喜欢**看？请举例。

——

4.5. 有哪些国外的娱乐节目您**不喜欢**看？请举例。

——

............................................................................................................

**5. 新闻节目**

5.1. 您看过多少国外的新闻节目？比如：CNN 新闻，BBC 新闻。

☐ 非常多 ☐ 比较多 ☐ 一般 ☐ 比较少 ☐ 非常少

5.2. 外国新闻节目中所反映的文化与中国文化相比，他们之间的文化差异有多大？您在何种程度上可以接受这些文化差异？

| | **差异大小**<br>非常小 → 非常大<br>1 — 2 — 3 — 4 — 5 | **接受程度**<br>完全能 → 完全不能<br>1 — 2 — 3 — 4 — 5 |
|---|---|---|
| 主持人的 外表（比如外貌，着装，化妆） | ☐ ☐ ☐ ☐ ☐ | ☐ ☐ ☐ ☐ ☐ |

| | | |
|---|---|---|
| 主持人的 个性（比如表情，动作） | ☐ ☐ ☐ ☐ ☐ | ☐ ☐ ☐ ☐ ☐ |
| 主持和评论的 主观性，客观性 | ☐ ☐ ☐ ☐ ☐ | ☐ ☐ ☐ ☐ ☐ |
| 主持和评论的 看待问题的出发点，角度 | ☐ ☐ ☐ ☐ ☐ | ☐ ☐ ☐ ☐ ☐ |
| 话题的 多样性 | ☐ ☐ ☐ ☐ ☐ | ☐ ☐ ☐ ☐ ☐ |
| 话题的 敏感性 | ☐ ☐ ☐ ☐ ☐ | ☐ ☐ ☐ ☐ ☐ |

5.3. 请您判断，您对以下国家或地区（除中国以外）的新闻节目的喜爱程度。

| | |
|---|---|
| 亚洲（比如日本，韩国，印度） | ☐ 非常喜欢 ☐ 比较喜欢 ☐ 一般 ☐ 不太喜欢 ☐ 很不喜欢 |
| 美洲（比如美国，加拿大） | ☐ 非常喜欢 ☐ 比较喜欢 ☐ 一般 ☐ 不太喜欢 ☐ 很不喜欢 |
| 欧洲（比如英国，德国，意大利） | ☐ 非常喜欢 ☐ 比较喜欢 ☐ 一般 ☐ 不太喜欢 ☐ 很不喜欢 |
| 其他国家 | ☐ 非常喜欢 ☐ 比较喜欢 ☐ 一般 ☐ 不太喜欢 ☐ 很不喜欢 |

5.4. 有哪些国外的新闻节目您**喜欢**看？请举例。

——

5.5. 有哪些国外的新闻节目您**不喜欢**看？请举例。

——

......................................................................................................

**6. 音乐节目**

6.1. 您看过多少国外的音乐节目？比如：MTV，维也纳新年音乐会。

☐ 非常多 ☐ 比较多 ☐ 一般 ☐ 比较少 ☐ 非常少

6.2. 外国音乐节目中所反映的文化与中国文化相比，他们之间的文化差异有多大？您在何种程度上可以接受这些文化差异？

| | **差异大小**<br>非常小 → 非常大<br>1 — 2 — 3 — 4 — 5 | **接受程度**<br>完全能 → 完全不能<br>1 — 2 — 3 — 4 — 5 |
|---|---|---|
| 音乐风格 | ☐ ☐ ☐ ☐ ☐ | ☐ ☐ ☐ ☐ ☐ |
| 音乐质量 | ☐ ☐ ☐ ☐ ☐ | ☐ ☐ ☐ ☐ ☐ |

6.3. 请您判断，您对以下国家或地区（除中国以外）的音乐节目的喜爱程度。

| | |
|---|---|
| 亚洲（比如日本，韩国，印度） | ☐ 非常喜欢 ☐ 比较喜欢 ☐ 一般 ☐ 不太喜欢 ☐ 很不喜欢 |
| 美洲（比如美国，加拿大） | ☐ 非常喜欢 ☐ 比较喜欢 ☐ 一般 ☐ 不太喜欢 ☐ 很不喜欢 |
| 欧洲（比如英国，德国，意大利） | ☐ 非常喜欢 ☐ 比较喜欢 ☐ 一般 ☐ 不太喜欢 ☐ 很不喜欢 |
| 其他国家 | ☐ 非常喜欢 ☐ 比较喜欢 ☐ 一般 ☐ 不太喜欢 ☐ 很不喜欢 |

6.4. 有哪些国外的音乐节目您**喜欢**看？请举例。

——

6.5. 有哪些国外的音乐节目您**不喜欢**看？请举例。

——

……………………………………………………………………………………

**7. 语言**

7.1. 以下哪种国外电视节目您觉得比较习惯？

☐ 配过音的外国节目

☐ 原声，有字幕的外国节目

☐ 原声，没有字幕的外国节目

☐ 无所谓是否配过音

7.2. 您觉得外国电视节目的原声和配音的差异有多大？您在何种程度上可以接受这些差异？

| | **差异大小**<br>非常小 → 非常大<br>1 — 2 — 3 — 4 — 5 | **接受程度**<br>完全能 → 完全不能<br>1 — 2 — 3 — 4 — 5 |
|---|---|---|
| 配音的 声音 | ☐ ☐ ☐ ☐ ☐ | ☐ ☐ ☐ ☐ ☐ |
| 配音的 情感 | ☐ ☐ ☐ ☐ ☐ | ☐ ☐ ☐ ☐ ☐ |
| 配音的 声音和画面的和谐性 | ☐ ☐ ☐ ☐ ☐ | ☐ ☐ ☐ ☐ ☐ |
| 翻译的 准确性 | ☐ ☐ ☐ ☐ ☐ | ☐ ☐ ☐ ☐ ☐ |
| 翻译中的 幽默 | ☐ ☐ ☐ ☐ ☐ | ☐ ☐ ☐ ☐ ☐ |

**8. 价值观**

8.1. 您觉得国外电视节目中所反映的价值观与中国的价值观相比，他们之间的文化差异有多大？您在何种程度上能接受这些差异？

| | 差异大小<br>非常小 → 非常大<br>1 — 2 — 3 — 4 — 5 | 接受程度<br>完全能 → 完全不能<br>1 — 2 — 3 — 4 — 5 |
|---|---|---|
| 个人主义 | ☐ ☐ ☐ ☐ ☐ | ☐ ☐ ☐ ☐ ☐ |
| 宗教观 | ☐ ☐ ☐ ☐ ☐ | ☐ ☐ ☐ ☐ ☐ |
| 道德观（比如道德标准） | ☐ ☐ ☐ ☐ ☐ | ☐ ☐ ☐ ☐ ☐ |
| 家庭观（比如以家庭为中心） | ☐ ☐ ☐ ☐ ☐ | ☐ ☐ ☐ ☐ ☐ |
| 自由，民主，平等 | ☐ ☐ ☐ ☐ ☐ | ☐ ☐ ☐ ☐ ☐ |
| 社会地位，竞争 | ☐ ☐ ☐ ☐ ☐ | ☐ ☐ ☐ ☐ ☐ |
| 爱国主义 | ☐ ☐ ☐ ☐ ☐ | ☐ ☐ ☐ ☐ ☐ |

**9. 生活方式**

9.1. 您觉得国外电视节目中所表现出来的生活方式与中国的生活方式相比，他们之间的文化差异有多大？您在何种程度上能接受这些差异？

| | 差异大小<br>非常小 → 非常大<br>1 — 2 — 3 — 4 — 5 | 接受程度<br>完全能 → 完全不能<br>1 — 2 — 3 — 4 — 5 |
|---|---|---|
| 生活习惯，风俗，仪式 | ☐ ☐ ☐ ☐ ☐ | ☐ ☐ ☐ ☐ ☐ |
| 生活环境 | ☐ ☐ ☐ ☐ ☐ | ☐ ☐ ☐ ☐ ☐ |

**10. 教育方法**

10.1. 您觉得国外电视节目中所表现出来的教育方法与中国的教育方法相比，他们之间的文化差异有多大？您在何种程度上能接受这些差异？

| | 差异大小<br>非常小 → 非常大 | 接受程度<br>完全能 → 完全不能 |
|---|---|---|

| | 1 — 2 — 3 — 4 — 5 | 1 — 2 — 3 — 4 — 5 |
|---|---|---|
| 自由性，独立性的培养 | ☐ ☐ ☐ ☐ ☐ | ☐ ☐ ☐ ☐ ☐ |
| 想象力，创造力的培养 | ☐ ☐ ☐ ☐ ☐ | ☐ ☐ ☐ ☐ ☐ |
| 长幼之间的平等性 | ☐ ☐ ☐ ☐ ☐ | ☐ ☐ ☐ ☐ ☐ |
| 对孩子的要求 | ☐ ☐ ☐ ☐ ☐ | ☐ ☐ ☐ ☐ ☐ |

---

**11. 总结**

11.1. 总体来说，您觉得国外电视节目中存在文化差异有多大？

☐ 非常小　☐ 比较小　☐ 一般　☐ 比较大　☐ 非常大

11.2. 总体来说，您在何种程度上可以接受这些差异？

☐ 完全能　☐ 大部分能　☐ 一半一半　☐ 小部分能　☐ 完全不能

---

**12. 个人信息**

12.1. 您的性别？

☐ 男　☐ 女

12.2. 您的年龄？

☐ 20 - 29 岁　☐ 30 - 39 岁　☐ 40 - 49 岁

12.3. 您现在主要居住在哪个城市？

☐ 北京　☐ 上海　☐ 广州　☐ 南京　☐ 西安

12.4. 您的最高学历是？

☐ 小学，初中或高中　☐ 专科或本科　☐ 硕士　☐ 博士

12.5. 您的平均月收入是？

☐ 0 - 999 元　☐ 1 000 - 4 999 元　☐ 5 000 - 9 999 元　☐ 高于 10 000 元

12.6. 您对外国文化了解多少？

☐ 很多　☐ 比较多　☐ 一般　☐ 比较少　☐ 非常少

| 请填写您的联系方式（自愿填写） | | | |
|---|---|---|---|
| 名字 | | QQ | |
| 电话 | | MSN | |
| 电子邮箱 | | Skype | |
| 地址 | | | |

**您已经成功完成了这份问卷!**

**非常感谢您的参与!**

# Appendix 6: Item Analysis

| | Perception index | | | | | Acceptance index | | | | |
|---|---|---|---|---|---|---|---|---|---|---|
| | Mis. | N | M | SD | SE | Mis. | N | M | SD | SE |
| (1) Movie & TV series | | | | | | | | | | |
| Appearance of character | .9% | 446 | 3.34 | 1.08 | .05 | 1.8% | 442 | 2.08 | 1.06 | .05 |
| Personality of character | 1.3% | 444 | 3.75 | 1.08 | .05 | 2.0% | 441 | 2.19 | 1.08 | .05 |
| Music style | 2.4% | 439 | 3.23 | 1.21 | .06 | 2.9% | 437 | 2.07 | 1.09 | .05 |
| Theme of drama plot | 2.0% | 441 | 3.60 | 1.13 | .05 | 2.7% | 438 | 2.13 | 1.05 | .05 |
| Poximity to daily life of d.p. | 3.3% | 435 | 3.22 | 1.11 | .05 | 4.4% | 430 | 2.29 | 1.16 | .06 |
| Fantasy of drama plot | 6.9% | 419 | 3.81 | 1.15 | .06 | 7.6% | 416 | 2.10 | 1.15 | .06 |
| (2) Sports program | | | | | | | | | | |
| Moderation/commentary | 2.2% | 440 | 3.34 | 1.13 | .05 | 4.2% | 431 | 2.30 | 1.06 | .05 |
| Sports culture | 3.8% | 433 | 3.28 | 1.13 | .05 | 5.1% | 427 | 2.28 | 1.10 | .05 |
| Commercialization | 6.2% | 422 | 3.51 | 1.13 | .06 | 8.2% | 413 | 2.48 | 1.19 | .06 |
| Sports types | 3.6% | 434 | 3.17 | 1.27 | .06 | 4.2% | 431 | 2.12 | 1.07 | .05 |
| (3) Documentary | | | | | | | | | | |
| Objectivity vs. subjectivity | 1.6% | 443 | 3.27 | 1.15 | .06 | 2.7% | 438 | 2.18 | .98 | .05 |
| Explicit vs. implicit | 2.0% | 441 | 3.26 | 1.10 | .05 | 2.9% | 437 | 2.15 | .95 | .05 |
| Starting point/perspective | 1.8% | 442 | 3.45 | 1.15 | .06 | 2.9% | 437 | 2.19 | 1.01 | .05 |
| Theme | 2.9% | 437 | 3.32 | 1.16 | .06 | 4.0% | 432 | 2.12 | 1.04 | .05 |
| (4) Entertainment show | | | | | | | | | | |
| Speaking style of presenter | 3.3% | 435 | 3.76 | 1.15 | .06 | 3.1% | 436 | 2.39 | 1.12 | .05 |
| Content of entertainment | 2.2% | 440 | 3.52 | 1.17 | .06 | 2.2% | 440 | 2.34 | 1.06 | .05 |
| (5) News program | | | | | | | | | | |
| Appearence of moderator | 3.6% | 434 | 3.13 | 1.14 | .06 | 4.9% | 428 | 2.22 | 1.02 | .05 |
| Personality of moderator | 3.8% | 433 | 3.35 | 1.06 | .05 | 4.4% | 430 | 2.29 | 1.01 | .05 |
| Objectivity vs. subjectivity | 4.2% | 431 | 3.25 | 1.15 | .06 | 6.7% | 420 | 2.33 | 1.03 | .05 |
| Starting point/perspective | 6.2% | 422 | 3.30 | 1.10 | .05 | 7.6% | 416 | 2.34 | 1.02 | .05 |
| Diversity of news theme | 5.8% | 424 | 3.36 | 1.13 | .06 | 7.8% | 415 | 2.29 | 1.03 | .05 |

| | | | | | | | | | |
|---|---|---|---|---|---|---|---|---|---|
| Sensitivity of news theme | 6.2% | 422 | 3.49 | 1.17 | .06 | 8.7% | 411 | 2.31 | 1.07 | .05 |
| (6) Music program | | | | | | | | | | |
| Music style | 2.4% | 439 | 3.56 | 1.17 | .06 | 3.8% | 433 | 2.07 | 1.03 | .05 |
| Music quality | 2.9% | 437 | 3.51 | 1.16 | .06 | 3.8% | 433 | 2.01 | 1.01 | .05 |
| (7) Language | | | | | | | | | | |
| Voice of dubbing | .7% | 447 | 3.55 | 1.21 | .06 | 1.1% | 445 | 2.71 | 1.20 | .06 |
| Emotion of dubbing | 1.3% | 444 | 3.54 | 1.12 | .05 | 1.6% | 443 | 2.67 | 1.17 | .06 |
| Coordination of dubbing | 1.1% | 445 | 3.42 | 1.12 | .05 | 1.3% | 444 | 2.72 | 1.18 | .06 |
| Acccuracy of translation | 1.6% | 443 | 3.49 | 1.08 | .05 | 2.7% | 438 | 2.76 | 1.18 | .06 |
| Humor in translation | 1.3% | 444 | 3.65 | 1.14 | .05 | 1.3% | 444 | 2.71 | 1.26 | .06 |
| (8) Value | | | | | | | | | | |
| Individualism | .9% | 446 | 3.65 | 1.09 | .05 | 1.8% | 442 | 2.27 | .99 | .05 |
| Religion | .9% | 446 | 3.74 | 1.12 | .05 | 2.0% | 441 | 2.51 | 1.12 | .05 |
| Morality | 1.6% | 443 | 3.45 | 1.08 | .05 | 2.0% | 441 | 2.39 | 1.06 | .05 |
| Family | .9% | 446 | 3.42 | 1.11 | .05 | 2.0% | 441 | 2.33 | 1.03 | .05 |
| Liberty/democracy/equality | 1.1% | 445 | 3.39 | 1.24 | .06 | 1.3% | 444 | 2.25 | 1.07 | .05 |
| Social status/competition | .7% | 447 | 3.57 | 1.16 | .06 | 1.6% | 443 | 2.35 | 1.02 | .05 |
| Patriotism | 1.3% | 444 | 3.13 | 1.23 | .06 | 2.2% | 440 | 2.26 | 1.07 | .05 |
| (9) Way of life | | | | | | | | | | |
| Habits/customs/rituals | 1.1% | 445 | 3.91 | 1.05 | .05 | 1.3% | 444 | 2.22 | 1.05 | .05 |
| Living environment | 1.8% | 442 | 3.83 | 1.09 | .05 | 2.0% | 441 | 2.19 | 1.03 | .05 |
| (10) Education method | | | | | | | | | | |
| Independence/freedoom | 1.1% | 445 | 3.90 | 1.18 | .06 | 1.1% | 445 | 2.10 | 1.03 | .05 |
| Imagination/creativity | 1.6% | 443 | 4.04 | 1.15 | .06 | 1.3% | 444 | 2.10 | 1.09 | .05 |
| Equality between young/old | 1.3% | 444 | 3.84 | 1.25 | .06 | 1.6% | 443 | 2.19 | 1.08 | .05 |
| Demand on children | 1.3% | 444 | 3.95 | 1.14 | .05 | 1.3% | 444 | 2.20 | 1.11 | .05 |

Note: SD = Standard Deviation, SE = Standard Error.